The Microsoft® Manual of Style for Technical Publications

Microsoft Corporation

Dana DeWitt Luther

PUBLISHED BY
Microsoft Press
A Division of Microsoft Corporation
One Microsoft Way
Redmond, Washington 98052-6399

Library of Congress Cataloging-in-Publication Data
Microsoft Manual of Style for Technical Publications / Microsoft
Corporation. -- 2nd ed.
p. cm.
Includes index.
ISBN 1-57231-890-2
1. Technical writing--Style--Handbooks, manuals, etc.
2. Microsoft Corporation. I. Microsoft Corporation.
T11.M467 1998
808'.066005--dc21 98-14000
CIP

Printed and bound in the United States of America.

4 5 6 7 8 9 MLML 3 2 1 0 9

Distributed in Canada by ITP Nelson, a division of Thomson Canada Limited.

A CIP catalogue record for this book is available from the British Library.

Microsoft Press books are available through booksellers and distributors worldwide. For further information about international editions, contact your local Microsoft Corporation office or contact Microsoft Press International directly at fax (425) 936-7329. Visit our Web site at mspress.microsoft.com.

Acquisitions Editor: Kim Fryer
Project Editor: Saul Candib

Contents

Introduction

Introduction to Version 3.0

This revised version of the *Microsoft Manual of Style for Technical Publications* (MSTP) is yet another evolution. The same principles apply to this version as to the previous versions: This guide is intended to give clear, up-to-date, and easy-to-use advice about usage and spelling of both general and computer-related terms, sentence style, technical writing issues, and design and interface elements.

The purpose of the revision is to add new terms and concepts, particularly those dealing with the Internet and the World Wide Web, to the extent that changing world stands still. It also includes more technical terms and suggests that some terms not acceptable to a general audience are acceptable or necessary in some technical documents. Finally, it corrects some errors.

As before, this guide is to help Microsoft writers and editors maintain consistency within and across products. It is not a set of rules. It does provide guidelines that have been discussed and reviewed by experienced writers and editors across the company. These guidelines represent their expertise and opinions of what best serves Microsoft documentation teams and their customers.

The *Microsoft Manual of Style* is not a text on technical writing or a guide to the different kinds of documentation that Microsoft produces. Nor does it cover all the terms specific to various Microsoft products, on the assumption that these are covered in project-specific style sheets.

New topics and other important changes

Many terminology additions, minor changes, and corrections appear throughout. Even if you think you know how a particular term is used, it's worthwhile to see if the guidelines have changed.

You should note two style changes. Small caps are no longer used for key names or for A.M. and P.M. Producing these in HTML was time-consuming and the gain in readability not sufficient to justify the work. Also, *KB* is now used as the abbreviation for *kilobyte,* rather than *K*. This change was made to reflect the more common usage around the world and to match the most frequent use in the interface.

There are also more substantial changes. The following topics contain heavily revised or new information in terms of usage, content, or policy and should be reviewed.

New or revised topics

-able, -ible
Accessible Documentation
applet
COM, ActiveX, and OLE Terminology
Gerunds
Headings and Subheadings
HTML
HTTP
Hyperlinks
Java, JScript, JavaScript
Key Names
Keywords and Online Index Entries
Screen Terminology
script, scripting language
search engine
setting
URL, address

Many new Internet terms are listed. They appear as individual topics in the MSTP and in the index. Jargon, slang, and Macintosh terms are also listed in the index as both separate entries and subentries under the conceptual topic.

In addition, the **list of acronyms and abbreviations** in Appendix A includes many new acronyms and corrections of previous ones. The separate appendix on words with numbers has been deleted because so many of the terms are obsolete. Useful terms from that appendix are still included in the guide.

The titles of usage and terminology topics are lowercase to reflect their spelling unless they should be spelled with a capital letter. Conceptual and procedural topics, on the other hand, are indicated by capital letters.

How to use this guide

Topics provide information ranging from a simple note on the correct spelling of a term to a thorough review of what to do, why, what to avoid, and what to do instead, with frequent correct and incorrect examples. Subentries to index keywords frequently refer to subsections of longer topics so that you can more easily find the information you want.

Topics that cover one subject or usage item are listed by the relevant term or title, such as "Anthropomorphism," "dialog," "Key Names," "Trademarks," and so on.

Topic titles listing two or more terms connected by a comma, such as "expand, collapse," reflect correct usage for two similar terms. And topic titles listing terms connected by *vs.,* such as "active vs. current" or "who vs. that," show that the two similar terms have different meanings in some contexts. Each term appears separately in the index.

If a term is followed by an abbreviation for a part of speech (*adj, n, v*), such as "drag-and-drop (adj)," avoid using the term in other ways. Terms not followed by a part-of-speech designation are either acceptable in all forms or clearly useful in only the form indicated.

There are two kinds of cross-references: *See,* which includes a reference to the primary topic where the information is covered, and *See also,* which points to related information. We have made an effort to minimize cross-references by often including enough information within a short topic to answer the most basic questions. Cross references appear in **bold type**. When a hyperlink is used in an example, it is <u>underlined</u>.

Most information presented here applies to all documentation, technical and end user. Terms specifically designated *technical,* however, are primarily acceptable in material intended for advanced users, programmers, and administrators, not for the average user. Terms suitable only for programming documentation are so indicated. The term *end user* is used here to differentiate the average user from the technical user.

Italic is used to refer to terms used as terms; examples of usage appear in quotation marks.

The comment *do not use* means just that; *avoid* means that you can use the term in special circumstances if no other word is accurate.

Other standard reference works

The following reference materials are the authorities for issues not covered in this guide:

- *American Heritage Dictionary of the English Language,* 3rd ed. Boston: Houghton Mifflin Company, 1992.
- *Chicago Manual of Style,* 14th ed. Chicago: The University of Chicago Press, 1993.
- *Harbrace College Handbook,* 12th ed. Fort Worth: Harcourt Brace College Publishers, 1994.
- *Microsoft Press Computer Dictionary,* 3rd ed. Redmond, WA: Microsoft Press, 1997
- *Windows Interface Guidelines for Software Design,* Redmond, WA: Microsoft Press, 1995.

Using the Companion CD

This book comes with a CD that includes the following software and book information:

- A powerful, searchable HTML version of the *Microsoft Manual of Style for Technical Publications*, Second Edition, that enables you to quickly locate specific information, such as a procedure or a definition, with only a click of the mouse
- A powerful, searchable HTML version of the *Microsoft Press Computer Dictionary*, Third Edition, that enables you to search for specific terms.
- A traditional Windows Help version of the *Microsoft Manual of Style for Technical Publications*, Second Edition.
- Microsoft Internet Explorer 4.01

Installing Microsoft Internet Explorer 4.01

The text is best viewed in Microsoft Internet Explorer 4.01. For this reason, a copy of Internet Explorer 4.01 is included on the CD.

To install Internet Explorer 4.01 from the CD, choose Run from the Start menu, and then type **d:\IE401\ie4setup.exe** in the Run dialog box. (If necessary, replace *d* with the letter of your CD-ROM drive; for example, use *f* if your CD-ROM drive is installed as drive F). Then follow the instructions for installation as they appear.

When you run Internet Explorer after installing it, you will see the Internet Connection Wizard. This wizard helps you set up an account with an Internet service provider or establish a connection to your current service provider. (You do not have to be connected to a service provider to use the files on the CD.)

About the Electronic Books

The CD-ROM includes electronic versions of The Microsoft Manual of Style for Technical Publications, Second Edition, and of the Microsoft Press Computer Dictionary, Third Edition. These powerful versions of the books offer full-text search that enables you to locate specific information, such as a procedure or a definition, with only a click of the mouse.

Internet Explorer 4.0 or later is required to view the electronic books. As part of the installation process, the setup programs will automatically install Microsoft Internet Explorer version 4.01 if it is not already installed on your system. Microsoft Internet Explorer 4.01 runs on Microsoft Windows 98, on Microsoft Windows 95, or on Microsoft Windows NT with Service Pack 3 installed.

If you're running Windows NT, you must install Service Pack 3 (not included) before you attempt to install the electronic books. For information about downloading the Microsoft Windows NT Service Pack 3, connect to http://backoffice.microsoft.com/downtrial/moreinfo/nt4sp3.asp.

To install the electronic version of the Microsoft Manual of Style for Technical Publications, Second Edition, do the following:

1. Insert the CD into your CD-ROM drive.
2. Click **Start** on the Windows taskbar.
3. Choose **Run** from the **Start** menu.
4. Type **d:\ebook\mstp\setup.exe** (where *d* is the letter of your CD-ROM drive letter).
5. Click **OK**.
6. Follow the setup instructions that appear.

To install the electronic version of the Microsoft Press Computer Dictionary, Third Edition, do the following.

1. Insert the CD into your CD-ROM drive.
2. Click **Start** on the Windows taskbar.
3. Choose **Run** from the **Start** menu.
4. Type **d:\ebook\mspcd\setup.exe** (where *d* is the letter of your CD-ROM drive letter).
5. Click **OK**.
6. Follow the setup instructions that appear.

The Setup programs for the electronic books install desktop icons and Start menu items identified with each title. If it does not already exist, the Setup programs create a Microsoft Press group for the items. To view the electronic books, you can either select from the Start menu or double-click one of the desktop icons.

Using the Windows Help Version of the Book

To use the traditional Windows Help version of the book, create a new folder called MSTP on your hard drive. Then copy the contents of the WinHelp folder on the CD to the new folder on your hard drive. To open the Help file, double-click the icon labeled MSTP.hlp.

Additional Information

Every effort has been made to ensure the accuracy of the book and the contents of this companion disc. Microsoft Press provides corrections for books through the World Wide Web at http://mspress.microsoft.com/mspress/support/

If you have comments, questions, or ideas regarding the book or this companion disc, please send them to Microsoft Press via e-mail at:

MSPINPUT@MICROSOFT.COM.

or via postal mail to:

Microsoft Press
Attention: Project Editor, Microsoft Manual of Style for Technical Publications, Second Edition
One Microsoft Way
Redmond, WA 98052-6399

Please note that product support is not offered through the above addresses.

For late-breaking information, look for the readme file on the companion CD.

Abbreviations and Acronyms

Use abbreviations (the shortened form of a word) and acronyms (words formed from the initial letters of a phrase) sparingly. Some usually acceptable abbreviations are those for bytes, A.M. and P.M., the United States (U.S.) and United Kingdom (U.K.), and, in some cases, units of measure.

For specific abbreviations and acronyms, check for individual entries in this guide; see the **List of Acronyms and Abbreviations** (Appendix A), or **Measurements**.

> **NOTE** Technically, an abbreviation is a shortened form of a word, an acronym is a pronounceable word, and an initialism is an abbreviation formed from the initial letters of words in a phrase, pronounced as individual letters (for example, SDK). In this guide, acronym is usually used to refer to an initialism.

In general, spell out the complete term the first time an abbreviation or acronym appears in the text, reference topic, or Help topic. Then show the abbreviation or acronym within parentheses.

Correct

1-gigabyte (GB) hard disk
information stored in random access memory (RAM)

In subsequent references, you can use just the abbreviation or acronym. However, use editorial judgment. In related online documents, spelling out the term once in the overview or a primary topic may be sufficient. In long printed documents, it may be a good idea to spell out the abbreviation or acronym again when it appears in a later chapter or if many pages separate subsequent references from the spelled-out term. On the other hand, if a common abbreviation or acronym such as "KB" or "RAM" appears often in various Help topics, for example, it's not necessary to always spell it out. Also, check with your localization specialist to see if an acronym is familiar to foreign users. If it is, you can spell it out less frequently.

It's acceptable to use an acronym in a heading, but do not spell out its meaning in the heading. Instead, use and spell out the full term in the first sentence after the heading, if it hasn't been spelled out previously.

Choose a preceding indefinite article ("a" or "an") based on the acronym's pronunciation—for example, "an ANSI character set" or "a WYSIWYG system." To form the plural of an acronym, use a lowercase "s" without an apostrophe.

Correct

a SCSI system
an SDK
several IFSs
an OEM
three OEMs

In general, do not capitalize a spelled-out phrase; see individual entries in your project style sheet and this guide for exceptions.

Correct

random access memory (RAM)
American National Standards Institute (ANSI)

-able, -ible (suffix)

If you can't find the spelling of a word ending in the suffixes "-able" or "-ible," in the *American Heritage Dictionary,* use these guidelines to form the spelling:

- Drop the final "-e" (the most usual practice in American English): "scalable," "updatable."
- For words ending in "-y" as a final syllable, change the "y" to"i": "undeniable."
- For words ending in "-ce" or "-ge," retain the final "e" to maintain the soft sound: "bridgeable," "changeable."
- For words that double the final consonant in past participle form, double the consonant before the suffix: "biddable," "forgettable." Exception: words ending in -fer (transferable).

The suffix "-able" is much more common than "-ible," especially for new word formations.

For more information, see *Fowler's Modern English Usage.*

abort

Do not use in end-user documentation; instead, use "end" to refer to communications and network connections, "quit" for programs, and "stop" for hardware operations.

"Abort" is acceptable to use in programmer or similar technical documentation if it is a function name, parameter name, or otherwise part of a name in the API, but avoid it otherwise. In general text, use another appropriate word instead.

Correct

To end your server connection, click Disconnect Network Drive on the Tools menu.
Quit all programs before you turn off your computer.
To stop a print job before it's finished, click Cancel.
The PHW_CANCEL_SRB routine is called when the minidriver should cancel a request with STATUS_CANCELLED.

above

Do not use to mean earlier in a book or online document; use *previous, preceding,* or *earlier* instead. You can also use *earlier* to refer to a chapter or section heading. Do not use *above* as an adjective preceding a noun, as in "the above section."

To show a cross-reference to another Web page, use a specific HTML hyperlink. Do not make assumptions about the user's path through a site. Even if you refer to a location on the same scrollable Web page, make the reference itself a link. Do not use *above.*

Correct

See What Is a Copyright?
See "Connecting to the Network," earlier in this chapter.

SEE ALSO **Cross-References**

accelerator key

Obsolete term to refer to refer to a keyboard key or key combination. Do not use. Use *shortcut key* instead.

SEE ALSO **shortcut key**

access

Acceptable to use in the sense of accessing data or a process, especially in programmer documentation. Otherwise, avoid as a verb, as in "access a program." It's technical jargon. Use *gain access to, log on to, start, switch to,* or another similar term instead.

Correct

Start the program either from the Start menu or from Windows Explorer.
You can access your personal data from the company intranet.
You can create shortcuts to quickly switch to programs you use often.

SEE ALSO **start, switch**

access key

The key that corresponds to an underlined letter on a menu, command, or dialog box option. Use this term only in material about customizing the interface. In nontechnical material, use *the underlined letter.*

Access keys and shortcut keys as they appear on the menu

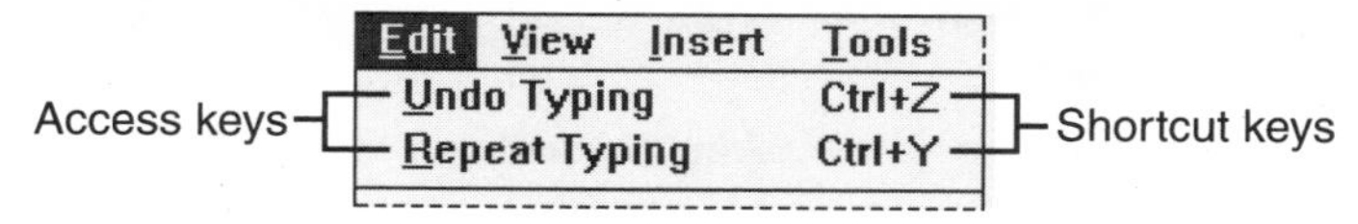

SEE ALSO **Key Names, shortcut key**

A B C D E F G H I J K L M N O P Q R S T U V W X Y Z

access privileges

The type of operations allowed a given user for certain system resources, but not specific files. These are generally allowed by the system administrator and help maintain system security. Contrast with **access rights**.

For more information, see the *Microsoft Computer Dictionary*.

access rights

The permissions of a user to view, enter, or modify a file, folder, or system. Not the same as access privileges, which refer to the permissions of a user to use only certain system resources, such as network or file servers.

accessible

Reserve *accessible* to refer to things that all people, including those with disabilities, can easily use. Don't use it to mean *simple*. Instead, use terms such as *easy to learn, easy to use, intuitive,* and *clear interface,* or refer to the specific characteristics that make something easy to use, such as *intelligent Help system.*

Correct

A range of enhancements makes multimedia products easier to install and use and provides a great platform for home entertainment.
You can customize the color schemes to make the program more accessible to people with disabilities.

Avoid

A range of enhancements makes multimedia products more accessible and provides a great platform for home entertainment.

Accessible Documentation

Documentation can be designed to maximize accessibility by using:

- Accepted standards for accessible Web pages
- Clear and concise writing.
- Graphics and design that help make documentation readable.
- Sensitive terminology.

Microsoft is publicly committed to making its products and services—including documentation—easy for everyone to use. To do so, products must accommodate a range of physical and cognitive abilities or different levels of familiarity with English.

For an introduction to and overview of Microsoft's progress in making products accessible and for ideas about what everyone can do to assist in these efforts, see the Microsoft Accessibility and Disabilities site at //www.microsoft.com/enable. The section "For Developers and Authors" is especially useful for guidelines.

Accessible Web pages

Keep in mind that not only do people with various kinds of disabilities have to get information from your Web site, but so do people using various kinds of browsers, or who have graphics turned off, and who may not have the latest technical wizardry. The guidelines given here are brief reminders. You can find more details about the rationale for these guidelines and some ways to accomplish them on these Web sites:

- Microsoft's guidelines for Accessible Web Pages at //www.microsoft.com/enable/dev/web.htm
- The Center for Information Technology Accommodation (CITA) of the U.S. General Services Administration at //www.gsa.gov/coca/WWWcode.htm
- Trace Research and Development Center at the University of Wisconsin at //trace.wisc.edu

The guidelines in the following list are based on those sources:

- Always provide **alt text** for graphic images. For simple images a brief but accurate description is enough. Use an asterisk (*) for bullets and leave blank information about invisible placeholders. For more complex images, provide a link to a separate page with more details.
- Provide text links in addition to image maps.
- Write link text that is meaningful but brief. Do not use phrases such as "click here." Use links that can stand alone in a list. Use the <TITLE> tag to distinguish links and names in image maps from ambiguous or duplicate text.
- Plan links and image map links so that navigation with the TAB key moves from left to right and top to bottom, not randomly.
- If you use frames, provide alternate pages without them.
- If you use tables, provide alternate pages without them. Make sure tables make sense when read from left to right, top to bottom. Note that Internet Explorer versions 3 and 4 have some different and strict requirements.
- Provide closed captions, transcripts, or descriptions of audio content.
- Avoid using scrolling marquees.
- Provide very simple alt text for images used as icons, such as "bullet" or "*," not "Little blue ball."

Accessible writing

Many of the following suggestions for maximizing accessibility also help make documentation clearer and more useful for everyone:

- Provide clear, concise descriptions of the product and initial setup, including a section or card that gets the user up and running with the basic features.
- Keep the number of steps in a procedure short. Individuals with cognitive impairments may have difficulty following multistep procedures. Research has shown that most procedures should be no more than seven steps, but fewer are better.
- Keep sentence structure simple. Try to limit most sentences to one clause. Individuals with language difficulties and those who speak English as a second language (including some people who are deaf) may have difficulty understanding complex text.

A B C D E F G H I J K L M N O P Q R S T U V W X Y Z

- Provide descriptions that do not require pictures, or provide both pictures and written descriptions. Using only diagrams causes difficulty transcribing to other media. To test whether the writing is effective, try removing, one at a time, first the words and then the pictures. With only one method, can you still figure out what to do?
- Avoid directional terms (left, right, up, down) if possible. Individuals with cognitive impairments may have difficulty interpreting them as do blind users relying on screen-reading software.
- Emphasize key information, and put it near the beginning of the text. Use bullets or headings to additionally emphasize important points.
- Keep paragraphs short or otherwise create small sections or text groupings.

Accessible graphics and design

It is possible to work within the requirements of Microsoft's standard design templates to make a product's written documentation as visually accessible as possible. For example, use short paragraphs and break up long passages of text with subheadings.

Follow these guidelines for visually accessible documents:

- Avoid color-coding. For example, rather than using color alone to convey that information is important to reading or understanding, use additional cues such as textual annotations or the underlines in Help. Or use patterns as well as colors to indicate different types of information in charts and graphs.
- Avoid hard-to-read color combinations, such as red and green or light green and white. People with some types of color blindness may have difficulty seeing the differences between the colors.
- Avoid screened art. Contrasting black and white is easiest to read. Avoid especially text on a screened background, which is difficult to see and for a machine to scan. For the same reason, avoid shaded backgrounds and watermarks or other images behind text.
- Avoid printing text outside a rectangular grid. People with low vision may have difficulty seeing text outside of an established grid. Try to keep text in a uniform space for both visibility and scannability.

Acceptable terminology

In general, say that "a person has a kind of disability," not "the disabled person." That is, consider the person first, not the label.

If necessary, use the following terms to describe people with disabilities or the disabilities themselves.

Use these terms	*Instead of*
Blind, has low vision	Sight-impaired, vision impaired
Deaf or hard-of-hearing	Hearing-impaired
Has limited dexterity, has motion disabilities	Crippled, lame
Without disabilities	Normal, able-bodied, healthy
One-handed, people who type with one hand	Single-handed
People with disabilities	The disabled, disabled people, people with handicaps, the handicapped
Cognitive disabilities, developmental disabilities	Slow learner, retarded, mentally handicapped
TTY/TDD (to refer to the telecommunication device)	TT/TTD

accounts receivable

Not *account receivables.*

acknowledgment

Do not spell with an *e* between the *g* and the *m.* For a section of a book acknowledging the contributions of other people, use the plural *acknowledgments* even if there's only one.

action bar

Do not use; use *menu bar* instead.

action button

Do not use; use *button* or *command button* instead.

activate

Avoid; use a term such as *open, start,* or *switch to,* depending on your meaning.

Active Voice vs. Passive Voice

In general, use the active voice, which tells who or what is performing the action of the sentence.

Avoid the passive voice except when necessary to avoid a wordy or awkward construction; when the subject is unknown or not the focus of the sentence; or in error messages and troubleshooters, when the user is the subject and might feel blamed for the error if the active voice were used. Passive voice is more often used and acceptable in programmer documentation.

Preferred (active voice)

You can divide your documents into as many sections as you want.
Data hiding provides a number of benefits.
Windows 95 includes many multimedia features.

Acceptable use of passive voice

The Include directive (#include) should appear in the header file for the fastest execution.
It is recommended that you choose the typical installation option.

Avoid (passive voice)

Your document can be divided into as many sections as you want.
A number of benefits are provided by data hiding.
Many multimedia features are included in Windows 95.

Use the active voice for column headings in tables that list user actions.

SEE ALSO **Error Messages, Tables, Verbs**

active vs. current

Use *active* or *open,* not *current,* to refer to open and operating windows, programs, documents, files, devices, or portions of the screen (such as an "open window" or "active cell"). But use *current* to refer to a drive, directory, folder, or other element that does not itself change.

> **NOTE** If active causes confusion with ActiveX, try to write around—for example, be as specific as possible in naming the active element.

Correct

Change the formula in the active cell.
To switch between open documents, on the Window menu click the document you want to switch to.
Windows Explorer indicates which folder is the current one.

ad hoc

Always two words. Means "established only for the specific purpose or case at hand."

adapter (adj, n)

Not *adaptor.*

add-in, add-on

Use *add-in* to refer to utility programs, drivers, and other software added to a primary program, such as Microsoft BookShelf in Word or the Analysis ToolPak in Microsoft Excel.

Use *add-on* to refer to a hardware device such as an expansion board or external peripheral equipment, such as a CD-ROM player, attached to the computer.

In end-user documentation especially, use these terms as adjectives: "the add-in program," "an add-on modem."

address

General term referring to the path to Internet and intranet sites and to e-mail user names and domains. Use *address* to refer to Internet addresses in most general and end-user material. If necessary for clarity, use a term such as "hyperlink" or "HTTP" to define "address."

Use *URL* (Uniform Resource Locator) in technical material.

Use *path* to refer to the hierarchical structure of an operating system from root directory through file names.

SEE ALSO **URL**

adjacent selection

Use instead of *contiguous selection* to refer to a multiple selection (of cells, for example) in which the items touch.

administer

Not *administrate.*

administrator

Use *system administrator* unless you must specify a particular kind of administrator, such as a network administrator or a database administrator.

Note spelling (ends in *or,* not *er).*

Administrator program

Not Administrator Program.

affect vs. effect

Double-check the use of these words in text, because the spellings are often confused and the spelling checker won't catch their misuse.

Affect is usually a verb meaning "to influence"—for example, "Deleting a link on the desktop does not affect the actual program." In psychology, *affect* can be a noun, but this usage is not needed in the software world—for example, "The patient's affect was inappropriate."

Effect is usually a noun meaning "a result"—for example, "The effect of the change was minimal."

Effect can also be a verb meaning "to cause something to happen," but this is a less common usage—for example, "Good software design can effect a change in users' perceptions."

Here's a handy way to remember the differences between the most common uses of these words—*affect* means to "cAuse," so it has an *a. Effect* means the "rEsult," so it has an *e.*

Correct

Late software can affect a schedule adversely.
One effect of late software is schedule slippage.

afterward

Not *afterwards.*

against

Do not use to refer to running a program on a particular platform or operating system. It is acceptable in technical and database documentation in the sense of evaluating a value against an expression or running a query against a database.

Correct

Show reference queries can be run against the Guide database.

Incorrect

If you want a program built against the newest version of DirectDraw to run against an older version, then define DIRECTDRAW_VERSION to be the earliest version of DirectDraw you want to run the program against.

alarm (n)

Avoid; use *beep* as a noun to refer to a sound, as in "when you hear the beep."

alert (n)

Avoid; use *message* instead. *Error message* is acceptable in technical documentation when necessary to differentiate types of messages.

align, aligned on

Use *align* instead of *justify* and *aligned on* instead of justified or *flush to. Right-aligned* and *left-aligned* are correct. It's acceptable to use a phrase such as "aligned with each other."

Correct

Align the text on the left.
The text is aligned on both the left and the right.
Align the text with the headings.

allow

Avoid *allow* in the sense of a program permitting a user to do something, especially in end-user documentation. Use *you can* if possible. However, you can use *allow* or *enable* in instances where you must use the third person

Correct

Microsoft Exchange allows a user to log on as a guest.
This function allows the program to print lists of files.
Using this function, you can print lists of files.

SEE ALSO **Anthropomorphism, can vs. may, enable**

alpha

Refers to the version of a software product that is completed and ready for internal testing. Alpha versions are usually not released to external beta testers.

SEE ALSO **beta**

alphabetical

Not *alphabetic.*

alphanumeric

As in “alphanumeric characters,” “alphanumeric mode.” Not *alphanumerical.*

alt text

The common term for the descriptive text that appears as an alternative to a graphic image on Web pages. The text is indicated in the HTML file by the attribute ALT. The code used for the graphic and the alt text looks like this:

```
<IMG SRC=“image.gif” ALT=“Add the description of the image here”>
```

Always use alt text whenever you use a graphic, and always briefly describe the graphic. Do not use a word such as “graphic” or “image” alone, but a description such as “Red rose” is acceptable. Many users turn graphics off, and screen readers cannot interpret pictures, so a description of the image is necessary.

A.M., P.M.

Use all caps and periods.

Do not use “A.M. and “P.M.” to refer to noon or midnight. This use is ambiguous.

Correct

The meeting is at 12:00 noon.
The date changes at exactly 12:00 midnight.

among vs. between

Use *between* when comparing two items or when denoting a one-to-one relationship, regardless of the number of items. Use *among* when the emphasis is on distribution rather than individual relationships.

Correct

Move between the two programs at the top of the list.
Switch between Windows-based programs.
You can share folders and printers among members of your workgroup.

ampersand (&)

Do not use *&* in text or headings to mean *and* unless you are specifically referencing the symbol on the interface.

The ampersand is also a special symbol in HTML that precedes the code name or number of a special character that a browser may not correctly display otherwise. For example, to show less-than (<) and greater-than (>) signs on a Web page you would use this HTML code:

```
&#60; &#62;
```

and/or

Avoid using this construction; choose either *and* or *or* or rewrite the sentence. If avoiding it makes a sentence long or cumbersome, however, it's okay to use it.

Correct

You can save the document under its current name or under a new name.
Will the new version contain information on how to write object-oriented code and/or use the class libraries?

Anthropomorphism

NOTE These guidelines do not apply to intentional anthropomorphism such as the wizards, assistants, characters, and guides built into various programs such as Publisher and Office.

In general, avoid giving hardware or software human characteristics or emotions. Occasionally, however, you can most accurately convey information by using anthropomorphic or figurative language. Programming documentation, in particular, lends itself to some anthropomorphic treatment. In all documentation it's acceptable that programs and commands do things, but avoid words that convey emotions (such as *refuses* or *wants),* behavior *(forces, tries),* and intellect *(knows, realizes)* rather than more straightforward actions.

Guidelines for evaluating anthropomorphism

To evaluate whether figurative language is effective, consider the following questions:

- Is the proposed anthropomorphic or figurative language the clearest way to convey meaning? Rewrite this language if it interferes with meaning or is ambiguous, inaccurate, or overextended.

  #### Acceptable

  The program denies access to users without the correct permissions.
  Mail prompts you to type your password.
  The design assistant guides you through the process of publishing a newsletter.

  #### Questionable

  The program won't let you in if you don't have the correct permissions.
  Mail will tell you to type your password.
  The design assistant will take you by the hand and lead you through the process of publishing a newsletter.

- Is the language typical of other documentation or writing on the subject?

  In object-oriented programming, for example, it is possible to talk about software entities that "do" and "know" things; however, use such language as a convenient shorthand only after establishing the figurative nature of the usage. It should not be used to explain the properties and behavior of complex software objects.

Correct

But unlike a Pascal record type, the button class also includes "methods." In Object Pascal, methods are procedures and functions. Thus, an object in a class contains both data and operations. A button class would have methods to display, hide, and dim a button; to test whether a button has been clicked; and so on. Think of an object in a button class as a capsule that "knows" how to behave the way a Macintosh button is supposed to.
You don't have to use sizeof to find the size of a Date object, because new can tell what size the object is.

Without suitable preparation to guide the reader, the same usage is questionable or unacceptable.

Questionable

Documents know how to manage data; views know how to display the data and accept operations on it.

Less figurative but still accurate

Documents manage data; views display the data and accept operations on it.

- Is the language appropriate to the audience and clear in context?

  In anthropomorphic products, it may be appropriate for a character to talk informally to the user. Likewise, in technical products, anthropomorphic shorthand such as "child threads" that "inherit" characteristics is widely used and accepted. Straining to avoid such use would likely destroy some of the clarity.

Words to watch out for

The following verbs may be acceptable in the right context, but they often signal excessive or inappropriate anthropomorphism. Some are appropriate only in technical documentation. This list is not exhaustive. Check your project style sheet.

allow	discard	prevent
answer	ignore	realize
assume	impersonate	recognize
behave	inherit	refuse
capture	interested in	remember
decide	know	think
demand	own	understand
deny	persist	wait

anti-aliasing

A technique for making jagged edges look smooth on the screen. Note the hyphen. For more information see the *Microsoft Press Computer Dictionary.*

Apostrophes

Use apostrophes to form the possessive case of nouns and to indicate a missing letter in a contraction.

Form the possessive case of a singular noun by adding an apostrophe and an *s,* even if the singular noun ends in *s, x,* or *z.*

Correct

insider's guide
Burns's poems
Berlioz's opera
an OEM's product
Microsoft's products

NOTE It is acceptable to use acronyms and company names in the possessive case.

Differentiate between the contraction *it's* (it is) and possessive pronoun *its*. Never use an apostrophe with possessive pronouns (not *your's*).

Do not use an apostrophe to indicate the plural of a singular noun (not *Microsoft's program's*).

SEE ALSO **Possessives**

appears, displays

Use *appears* as an intransitive verb; use *displays* as a transitive verb. If necessary in context, you can use the passive *is displayed.*

Correct

If you try to quit the program without saving the file, a message appears.
The screen displays a message if you don't log on accurately.
A message is displayed if you don't log on accurately.

Incorrect

If you try to quit the program without saving the file, a message displays.

appendix (s), appendixes (pl)

Do not use *appendices* for the plural.

Apple menu

Macintosh only. Refers to the menu identified by the Apple Computer logo. Use both the name and the symbol on first reference, the name alone thereafter. It takes the article *the,* as in "the Apple menu."

Apple menu icon

applet

In current usage "applet" refers to an HTML-based program that a browser downloads temporarily to a user's hard disk. It is most often associated with Java. Use it only to refer to Java-based applets.

Do not use *applet* to refer to a small Windows program or accessory. Instead use a more specific term, such as *program, add-in, utility*, or just the name of the program.

application

Avoid in end-user documentation. Use *program* instead. Acceptable in technical documentation, especially to refer to a grouping of software that includes both executable files and other components.

Do not use "application program."

SEE ALSO **applet, program vs. application**

application developer

Not *applications developer.* But *developer* or *programmer* is preferable.

application file

Do not use; use *program file* instead.

application icon

Avoid; instead, use the specific product name such as "the Word icon" whenever possible.

Application icon

Microsoft Word

application window

Avoid; instead, use the specific product name such as "the Word window" whenever possible.

arabic numerals

Use lowercase *a* for the word *arabic* when referring to numbers.

argument vs. parameter

An *argument* typically is a value or expression containing data or code that is used with an operator or passed to a subprogram.

A *parameter* is a value given to a variable until the operation is completed and is treated by the computer as a constant. Parameters are often used to customize a program for a particular purpose. For example, a date could be a required parameter in a function in a scheduling program.

In programmer or other technical documentation use the same term consistently to refer to the same kind of element.

The difference between an argument and a parameter is unimportant for end users because it appears so infrequently. In general, use *argument* in end-user documentation. Differentiate between the two only if necessary.

SEE ALSO **Command Syntax, Document Conventions**

arrow

In documentation for novice users, you may want to use *arrow* to identify the arrow next to a list box label. Do not use *up arrow* or *down arrow,* which refer to the **arrow keys** on the keyboard.

Correct

Click the Font arrow to display the list.

arrow keys

Use to collectively refer to the LEFT ARROW, RIGHT ARROW, UP ARROW, and DOWN ARROW keys on the keyboard. Specify "on the numeric keypad" if you need to differentiate.

Do not use *direction keys, directional keys,* or *movement keys.*

Use specific names to refer to other navigational keys, such as PAGE UP, PAGE DOWN, HOME, and END.

SEE ALSO **Key Names**

arrow pointer

Do not use; use *pointer* instead. See **pointer**.

Art, Captions, and Callouts

The appearance and placement of art (screen shots, built art, and conceptual art) in documentation are determined by the design. Consult your designer or the template information about your design for guidelines on the placement and appearance of art, captions, and callouts.

Content

Generally, in printed material, information that appears in captions and callouts should also appear in text, unless there's no possibility of misreading or confusion and the art appears in the immediate textual context. However, online material and some printed documents call for essential or procedural information to be placed solely in captions and callouts. This is acceptable, but note that doing so can affect accessibility.

1 Double-click My Documents. Find the file or folder you want to send, and then click it.

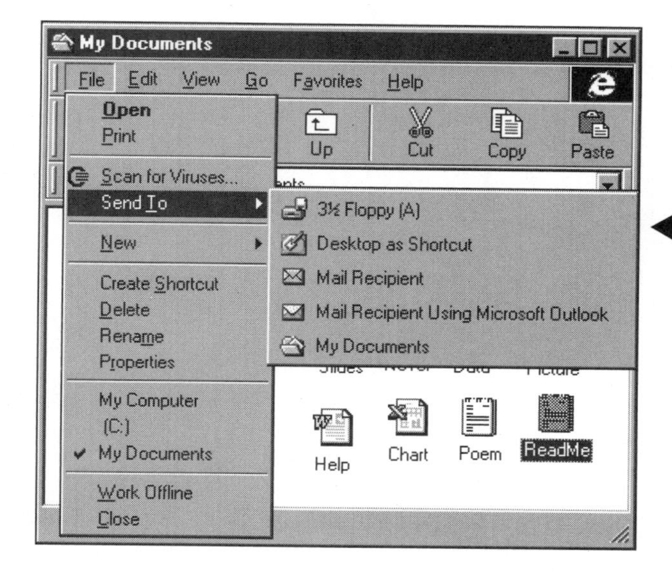

2 On the File menu, point to Send To, and then click where you want to send the file or folder.

Not all art needs captions or callouts, even within the same document. Use editorial judgment or see the project style sheet.

Always provide a brief description (see **alt text**) or a caption for a graphic on a Web page.

Cross-references to art

Avoid making cross-references to untitled or unnumbered art unless the relation of the art to the text is not immediately apparent or the content of the art furthers the explanation found in the text. In that case, a reference to the art is helpful. However, the art should appear as close to the reference as possible; so that you can write something like "The following illustration shows. ..."

In both online and printed material, avoid making any cross-references to art unless the art immediately precedes or follows the reference within the Help topic or on the page.

Use *preceding* and *following* (or more specific references) when referring to art, not *above, below, earlier,* or *later.* End the introductory sentence with a period, not a colon.

Correct

The following illustration shows file sharing with user-level access control.

Captions

There are two kinds of captions: title and descriptive. It's acceptable to use both kinds within one document.

Title captions label a piece of art and should be concise. Use sentence cap style and no end punctuation. Some teams use numbered titles for art. In that case, the pieces are referred to as "figures"—for example, "Figure 7.1 Arcs."

Figure 10-1 *The Internet Explorer Properties dialog box*

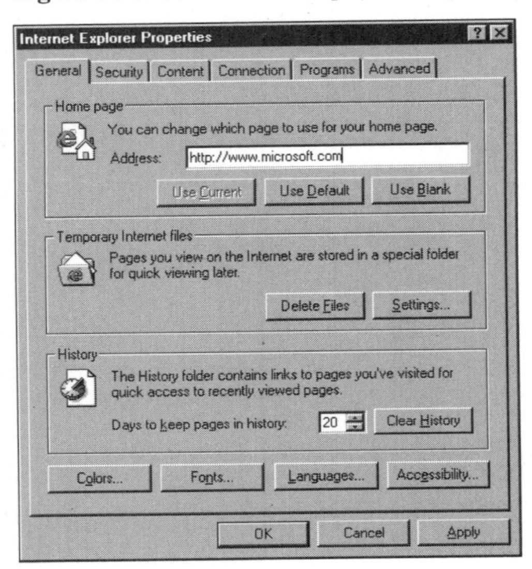

Descriptive captions explain something about the art but do not necessarily point to or call out anything in particular about the art. Writers and editors determine how to write descriptive captions and when to use them. Use sentence cap style for a descriptive caption, and end the caption with a period only if it is a complete sentence or a mixture of fragments and sentences.

Flying Windows screen saver, just one of the screen savers included with Windows 98

Callouts

A callout points to a specific item that the reader should notice in an illustration. Observe the following rules when writing callouts:

- Capitalize each callout, except those with an ellipsis.
- End the callout with a period only if the callout is a complete sentence.
- Avoid mixing fragments and complete sentences as callouts for the same piece of art. If you must, use end punctuation appropriate to each callout.

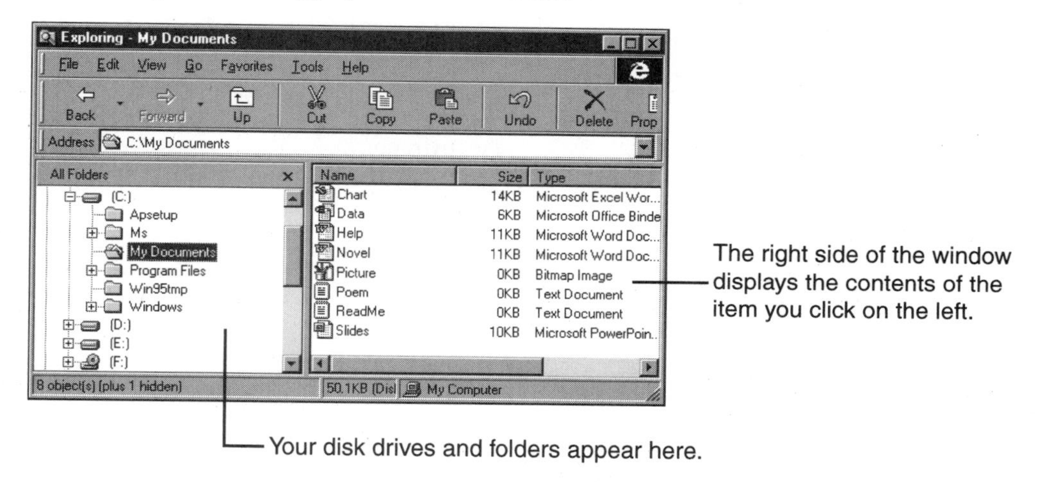

- For multiple-part callouts with ellipses, use lowercase for the first word in the second part of the callout. Leave one space before or after the ellipses, as shown in the example.

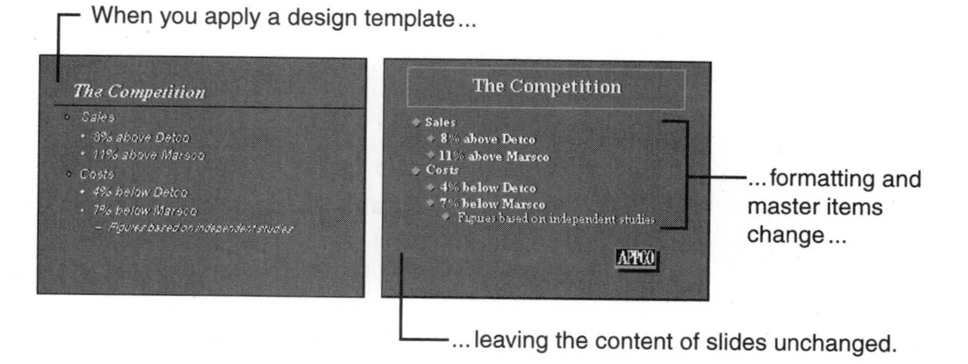

SEE ALSO **Capitalization, Headings and Subheadings, International Considerations**

A
B
C
D
E
F
G
H
I
J
K
L
M
N
O
P
Q
R
S
T
U
V
W
X
Y
Z

article

Use *article* to refer to topics in encyclopedias (such as the Microsoft Encarta multimedia encyclopedia) and similar reference programs and the contents of magazines, journals, and newspapers and newscasts, whether online or in print. For example, you can refer to a column on MSNBC or a product-related white paper as an "article." Do not use to refer to Help topics or sections or chapters of printed or online books.

as

Use as a conjunction meaning "in the same manner." Do not use as a synonym for *because* or *while* in subordinate clauses.

Correct

You can use the Forms Designer as a complete development environment.
Use the active voice whenever possible, because it's easier to translate.
Fill out your registration card while you wait for Setup to finish.

Incorrect

Use the active voice whenever possible, as it's easier to translate.
Fill out your registration card as you wait for Setup to finish.

as well as

Use in the sense of "in addition to," not as a synonym for *and.* But because *as well as* can also be used adverbially (as in "She can play as well as he"), be aware that its use may be confusing for localizers. If it's not important stylistically to use *as well as*, prefer *in addition to*.

Correct

With Word you can format whole documents, insert headers and footers, and develop an index, as well as write a simple letter.
With Word you can format whole documents, insert headers and footers, and develop an index, in addition to writing a simple letter.

assembly language (n), assembly-language (adj)

Not *assembler,* and not *machine language.* A low-level language that uses an assembler to translate the code into machine code.

assistive (adj)

Acceptable to use to refer to devices and organizations that help people with disabilities.

SEE ALSO **Accessible Documentation**

asterisk (*)

Not *star,* except when referring to the key on a telephone keypad. Indicates multiplication in programming languages and also serves as a **wildcard character** representing one or more characters.

at sign (@)

In Internet e-mail addresses, @ separates the user name from the domain name, as in username@hostname.com. It's pronounced "at" when spoken.

attribute

A characteristic that can change something. *Property* is more often used in the interface.

In HTML, an attribute is a named value within a tagged element that can change default characteristics of the tag. For example, in a table, the attributes WIDTH and HEIGHT specify the size of a table or table cells. The code for an HTML attribute looks like this:

```
<TABLE WIDTH=50% HEIGHT=50%>
```

In MS-DOS, files can have attributes such as hidden, read-only, and so on.

SEE ALSO **properties**

audit trail

Not *audit log.*

author (v)

Do not use as a synonym for *write.* It's unnecessary jargon.

auto (prefix)

In general, do not hyphenate words beginning with *auto,* such as *autoanswer, autodemo,* and *autodial,* unless it's necessary to avoid confusion. If in doubt, check *American Heritage Dictionary* or your project style sheet.

back end (n), back-end (adj)

Avoid in documentation; it's jargon. Instead, use a more specific term such as *server, operating system,* or *network.*

back up (v), backup (adj, n)

Note the difference in spelling, depending on use.

Correct

Back up the files before you turn off the computer, and save the backup on a floppy disk.

backbone (n)

Usually, a large, fast network connecting other networks. Acceptable to use in network documentation without defining. For more information, see the *Microsoft Press Computer Dictionary.*

BackOffice

An integrated family of server software for developing client/server applications. Note capitalization. Registered trademark.

backspace

Okay to use as a verb.

backtab

Do not use. If necessary to explain the procedure, refer to the SHIFT+TAB key combination.

backward

Not *backwards.*

base line vs. baseline

Use *baseline* (one word) to refer to an established standard, as in "baseline data." Use *base line* (two words) only to refer to the bottom alignment of capital letters in print (a typographic term).

baud

Refers to the rate of signals transmitted per second. Because baud is a rate, the phrase "baud rate" is redundant and unnecessary, although widely used.

Baud and bits per second are not necessarily the same, so do not use "bits per second" or "bps" as a synonym for baud. Modems are conventionally designated by bits per second or kilobits per second. A 28.8 Kbps modem runs at a different baud, depending on how events are coded for transmission.

When designating baud, use commas when the number has five (not four) or more digits.

because vs. since

Avoid using *since* in the sense of *because;* it's ambiguous. Use *because* to refer to a reason, *since* to refer to a passage of time. *Since* is often clearer when used with a gerund rather than a participle.

Correct

Because I installed the fast modem, I can download messages very quickly.
Since installing the fast modem, I can download messages very quickly.

Ambiguous

Since I installed the fast modem, downloading messages takes much less time than it did.

beep (n)

Use instead of *alarm* or *tone* to refer to a sound, as in "when you hear the beep."

below

Do not use to mean later in a book or Help topic; use *later* instead.

SEE ALSO **Cross-References**

beta

A software product ready to release for outside testing.

Marketing sometimes calls beta versions "preview" versions or something similar. Avoid such synonyms in documentation because the terms can be ambiguous.

bi (prefix)

In general, do not hyphenate words beginning with *bi,* such as *bidirectional, bimodal,* and *bimonthly,* unless it's necessary to avoid confusion. If in doubt, check *American Heritage Dictionary* or your project style sheet.

Bias-Free Communication

Microsoft supports, by policy and practice, the elimination of bias in both written and visual communication. Thus, documentation and art should show diverse individuals from all walks of life participating fully in various activities. Specifically, avoid terms that may show bias with regard to gender, race, culture, ability, age, sexual orientation, or socioeconomic class.

Use the following sections to evaluate your work and eliminate bias and stereotypes from it.

Avoid racial, cultural, sexual, and other stereotypes

- Use gender-neutral or all-inclusive terms to refer to human beings, rather than terms using *man* and similar masculine terms.

Use these terms	*Instead of*
Chair, moderator	Chairman
Humanity, people, humankind	Man, mankind
Operates, staffs	Mans
Sales representative	Salesman
Synthetic, manufactured	Man-made
Workforce, staff, personnel	Manpower

- Avoid the generic masculine pronoun. Use *the* instead of *his,* or rewrite material in the second person *(you)* or in the plural. Use *his or her* if you can do so infrequently and if nothing else works. However, do not use a plural pronoun such as *they* or *their* with a singular antecedent such as *everyone.*

Correct

A user can change the default settings.
You can change the default settings.

Incorrect

A user can change his default settings.
Everyone can change their default settings.
Each employee can arrive when he wishes.

- Use a variety of first names, both male and female, that reflect different cultural backgrounds.
- In art, show men and women of all ages, members of all ethnic groups, and people with disabilities in a wide variety of professions, educational settings, locales, and economic settings.
- Avoid stereotypes relating to family structure, leisure activities, and purchasing power. If you show various family groupings, consider showing nontraditional and extended families.
- Try to avoid topics that reflect primarily a Western, affluent lifestyle.

SEE ALSO **International Considerations**

Avoid stereotypes of people with disabilities

Not only should Microsoft products and documentation be accessible to all, regardless of disabilities, but documentation should positively portray people with disabilities.

- Avoid equating people with their disabilities. In general, focus on the person, not the disability. Whenever possible, use terms that refer to physical differences as nouns rather than adjectives. For example, use wording such as "customers who are blind or have low vision" and "users with limited dexterity."

  NOTE The phrases "she is blind" and "he is deaf" are acceptable.

- Do not use terms that depersonalize and group people as if they were identical, such as "the blind" and "the deaf."

Correct

Customers who are blind can use these features.

Incorrect

The blind can use these features.

- Avoid using terms that engender discomfort, pity, or guilt, such as *suffers from, stricken with,* or *afflicted by.*
- Avoid mentioning a disability unless it is pertinent.

Correct

Play-goers who are deaf or hard-of-hearing can attend signed performances.

Incorrect

Theaters now offer signed performances for the deaf. [depersonalizes people who are deaf and irrelevant to mention who the performances are for]

- Include people with disabilities in art and illustrations, showing them integrated in an unremarkable way with other members of society. In drawings of buildings and blueprints, show ramps for wheelchair accessibility.

SEE ALSO **Accessible Documentation**

For more information

For background reading and in-depth information, see the following sources:

Dumond, Val. *The Elements of Nonsexist Usage: A Guide to Inclusive Spoken and Written English.* New York: Prentice Hall Press, 1990.

Guidelines for Bias-Free Publishing. New York: McGraw-Hill, n.d.

Maggio, Rosalie. *The Bias-Free Word Finder: A Dictionary of Nondiscriminatory Language.* Boston: Beacon Press, 1991.

Schwartz, Marilyn. *Guidelines for Bias-Free Writing.* Bloomington, Indiana: University Press, 1995.

A B C D E F G H I J K L M N O P Q R S T U V W X Y Z

Bibliographies

One purpose of listing sources of information is to help readers go directly to that source. If you need to cite a source in documentation, follow the examples listed here.

Citing books and printed articles

To cite printed works, follow *The Chicago Manual of Style* "Documentation One" format. (Exception: Use the Postal Service abbreviation for states, not the abbreviations used by *The Chicago Manual.*) Bibliographies usually are formatted with a hanging indent, but that may not be possible in some Microsoft design templates. In that case, separate each entry with a line of space.

Only basic kinds of books and articles are listed here. For more information, see *The Chicago Manual of Style*, Chapter 15.

Books, general bibliographic style

The following paragraph lists the order and punctuation for each element in the citation of a book.

Author's name (last name first for the first author, first name first for additional authors). *Title, including subtitle.* Any additional information about the work, including editor's or translator's name and volume number. Edition number, if not the first. Facts of publication: Place, publisher, date.

Books, examples

Dupre, Lyn. *Bugs in Writing: A Guide to Debugging Your Prose*. Reading, MA: Addison-Wesley Publishing Co., 1995.

Li, Xia, and Nancy B. Crane. *Electronic Styles: A Handbook for Citing Electronic Information*. Rev. ed. Medford, NJ, Information Today, 1996.

Printed articles, general bibliographic style

Author's name. "Title of article." *Title of Periodical.* Volume and issue number (for journals only), date by month, day, year, page numbers.

The order of information and punctuation for the date differs between journals and popular magazines. For more information, see *The Chicago Manual of Style,* Chapter 15.

Printed magazine and journal articles, examples

Rosenthal, Marshal M. "Digital Cash: The Choices Are Growing." *Websmith*, May 1996, 6–9.

Vijayan, Jaikumar, and Mindy Blodgett. "Train Wreck at DEC." *Computerworld,* July 8, 1996, 1, 15.

Earle, Ralph, Robert Berry, and Michelle Corbin Nichols. "Indexing Online Information." Technical Communication: Journal of the Society for Technical Information 43 (May 1996): 146–56.

Citing electronic information

References to electronic information have the same intent and a format similar to the citations of printed material. That is, they follow the same general order of information such as author and title, but that information is followed by information such as the commercial supplier (if from an information service), the medium of availability (such as CD-ROM) or the Internet address, and the date accessed, if relevant. The important thing is to give enough information so that a user can find the source. Use lowercase for e-mail or other login names or follow the protocol of the e-mail service provider.

If the source appears both online and in print, give enough information so it can be found at either place. Rather than indicating page numbers of a magazine article that appears online, give an approximate length indication, usually in number of paragraphs.

This information is adapted from the electronic reference (//www.uvm.edu/~ncrane/estyles) to *Electronic Style*, cited in both the examples for books and the examples for Internet sites. *Electronic Style* itself follows MLA style rather than *The Chicago Manual of Style*, but the kind of information to cite is accurate.

CD-ROMs and computer programs, examples

"Washington." Encarta 98 Desk Encyclopedia. 1997. CD-ROM. Microsoft Corporation, Redmond, WA.

Visual Basic Ver. 4.0. Microsoft Corporation, Redmond, WA.

NOTE You do not need to cite a date of access for CD-ROMs and similar media.

Internet sites, examples

Schwartz, Delmore. "Survey of Our National Phenomena." The *New York Times Magazine*. April 15, 1956. Reprinted in *The New York Times Magazine*. April 21, 1996: 17 pars. Available http://www.nytimes.com/specials/magazine/titles.html.

UVM Bailey/Howe Library Reference Services. "Modern Language Association (MLA) Embellished Style." *Bibliographic Formats for Citing Electronic Information.* n.d.: 4 pp. Available http://www.uvm.edu/~ncrane/estyles/mla.html.

Discussion list messages and e-mail, examples

rrecome. "Top Ten Rules of Film Criticism." Online newsgroup posting. Discussions of All Forms of Cinema. Available e-mail: listserv@american.edu.Get cinema-l log9504A. August. 1995.

Someone, Jack (someone@microsoft.com). "New Terminology." E-mail to Nancy Someone (someonen@microsoft.com). March 5, 1996.

big-endian, little-endian (adj)

Big-endian refers to the method of storing numbers so that the most significant byte is placed first. *Little-endian* is the opposite. For more information, see the *Microsoft Press Computer Dictionary.* These are acceptable terms in programmer documentation.

bitmap (adj, n)

One word. Refers to a specific file format for online art. Do not use generically to refer to any graphic. Use *figure, picture,* or a similar term instead.

A B C D E F G H I J K L M N O P Q R S T U V W X Y Z

bitplane

One word. Refers to one of a set of bitmaps that together make up a color image. For more information, see the *Microsoft Press Computer Dictionary.*

bits per second

In general, spell out at first mention; then use the abbreviation *bps.* If you are sure your audience is familiar with the term, you do not need to spell it out.

Do not use as a synonym for *baud.*

SEE ALSO **baud**

bitwise

One word. Use in technical documentation only.

blank (adj, n)

Do not use as a verb.

board

Use to refer to a motherboard, system board, and similar hardware items. Otherwise, avoid; use *card* instead, as in *sound card* or *network interface card.*

bold (adj)

Not *bolded, boldface,* or *boldfaced.* Do not use as a verb.

Correct

To make the selected characters bold ...

Incorrect

To bold the selected characters ...

> **NOTE** The opposite of bold type is *roman,* although it's sometimes referred to as *light* or *lightface.*

SEE ALSO **Document Conventions, font and font style**

bookmark

In general Internet usage, and particularly in Netscape Navigator, a saved reference in the form of a URL or link to a particular location, page, or site, making it easy to return to. Use *favorite* to refer to a bookmark in Microsoft Internet Explorer.

SEE ALSO **favorite**

Boolean

Always capitalize.

boot

Avoid in end-user documentation; instead use *start, restart,* or *switch on* to refer to turning on the machine. Acceptable in technical documentation.

bootable disk

Avoid; use *system disk* or *startup disk* instead. It's acceptable to use *boot disk* in programmer documentation.

bot

Short for *robot.* Technical jargon, but a commonly used term on the Internet to refer to a program that performs a repetitive task, particularly posting messages to newsgroups and keeping Internet Relay Chat (IRC) channels open.

Avoid in most instances except in material about IRCs, chat rooms, and multiuser dungeons (MUDs), where it may be appropriate. Substitute a clearer and more descriptive term.

SEE ALSO **spider**

bottom left, bottom right

Avoid; use *lower left* and *lower right* instead, which are hyphenated as adjectives.

bounding outline

Technical term for the visible element (usually a dotted rectangle) that appears when a user selects a range of items. Also referred to as a *marquee* (do not use). It's acceptable to use *dotted rectangle* or *dotted box* if necessary to describe it, especially in end-user material, but in general use *bounding outline.*

box

Use instead of *field* in a dialog box to refer to any box, except a check box (for which use the complete term). For dialog box elements that display a list, such as a drop-down list box, you can use *list* rather than *box* for clarity.

Correct

the Read-Only box
the File Name box
the Hidden Text check box
the Wallpaper list

Incorrect

the User Name field

SEE ALSO **Dialog Boxes and Property Sheets**

bps

Acceptable abbreviation for *bits per second,* after it has been spelled out at first mention.

SEE ALSO **baud**

breakpoint

One word. Technical term related to testing and debugging.

Briefcase

A Windows 95 program. Do not precede with *the* or a possessive pronoun such as *your.*

Correct

In Windows 95, Briefcase is a program that helps you keep versions of files in sync between portable and desktop computers.

broadcast (adj, v)

A broadcast client and broadcast server are part of Broadcast Architecture, a Microsoft software and hardware design that enables personal computers to serve as clients of broadband digital and analog broadcast networks. Broadcast programming can include television, audio, World Wide Web pages, and computer data content.

The past tense of the verb *broadcast* is *broadcast,* not *broadcasted.*

browse

To scan Internet sites or other files, either for a particular item or for general items of interest. It's acceptable to use "browse the Internet," but use "browse through a list" (database, document, and so on), not "browse a list." Compare with **surf**.

browser

A program that interprets the HTML files posted on the Web, formats them, and displays them to the user. Microsoft's browser is Internet Explorer.

Use *browser*, not *viewer*, to refer to the interface to the Internet.

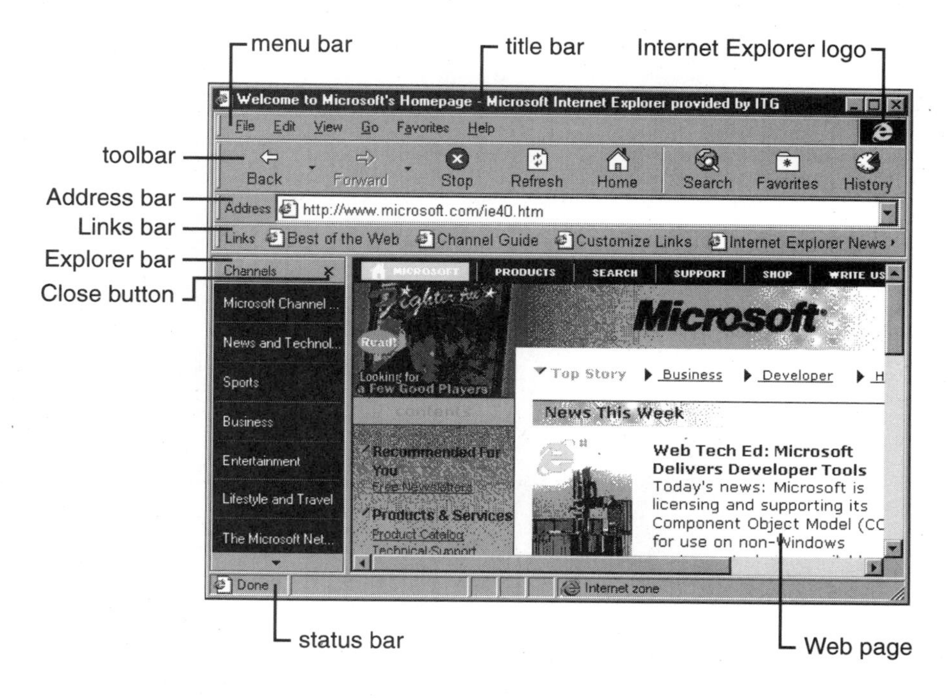

SEE ALSO **Screen Terminology**

bug

Acceptable to use without definition or apology for a software or hardware error.

build

Avoid in end-user documentation; use *create* for documents, charts, worksheets, and so on. Okay to use as a verb in programmer documentation to mean to compile and link code. Avoid using as a noun; it's jargon.

bulleted (adj)

Use "bulleted list," not *bullet list.* However, to refer to the typographic symbol (•) or a single item, as in a "bullet point," *bullet* is correct.

button

Lowercase, as in "Maximize button." Not *action button.* Use as the shortened form of *command button.*

SEE ALSO **Dialog Boxes and Property Sheets**

A B C D E F G H I J K L M N O P Q R S T U V W X Y Z

C

A programming language. Hyphenate as an adjective, as in "a C-language program."

Avoid using *C++* as an adjective, however. Instead, say something like "a program written in C++."

cable (n)

Do not use as a verb, as in "the printer is cabled to the computer." Use *connect* or a similar term instead.

cabling (n)

Avoid; use *cable* or *cables* instead. If necessary (as in some technical material about network connections), you can use *cabling* to refer to the combination of cables, connectors, and terminators.

cache vs. disk cache

Differentiate between these two terms. A *cache* generally refers to a special memory subsystem where data values are duplicated for quick access. A disk *cache* refers to a portion of RAM that temporarily stores information read from disk. See the *Microsoft Press Computer Dictionary.*

Do not use to refer to temporary Internet files that get downloaded to your computer as you browse the Web. It's jargon. The Internet Explorer term is Temporary Internet Files folder.

calendar (n)

Do not use as a verb; instead, use *schedule, list,* or another appropriate verb.

call out (v), callout (adj, n)

Two words as a verb; one word as a noun or adjective.

Correct

You should call out special features in the interface.
Add callouts to the art. The callout wording should be brief.

For information about using callouts, see **Art, Captions, and Callouts**.

caller ID

Okay to use *ID* without spelling out in this instance.

can vs. may

Use the verb *can* to describe actions or tasks that the user or program is able to do. Use *may* or *might* only to express possibility or when the result of an action is unknown or variable, not to imply that the user has permission to do something.

Correct

You can use the /b option to force a black-and-white screen display.
If you use the /b option, your code may not be portable.
If the table overlaps the text or the margin, you may need to resize the table and wrap text around it.
If the table overlaps the text or the margin, you can resize the table and wrap text around it.
Many new programs might run very slowly on less powerful computers.

Incorrect

You may use the /b option to force a black-and-white display.

In general, avoid *could;* it's seldom necessary. *Might,* however, connotes a sense of "a possible occurrence" without the suggestion of permission or recommendation, so it can be useful in some instances when *may* seems to imply permission.

Cancel button

In procedures, it's not necessary to use the words *the* and *button* with the name; use "click Cancel."

Cancel button

cancel the selection

Not *deselect* or *unmark.* Use *clear* to refer to check boxes.

canceled, canceling

One *l.* Do not use *cancelled* or *cancelling,* but use *cancellation* as a noun.

Capitalization

In general, use standard capitalization rules whenever possible—for example, common nouns are usually all lowercase and proper nouns are always capitalized. For most uses, follow these guidelines:

- Never use all uppercase letters for emphasis.
- Follow the capitalization rules or conventions of software or a specific product as necessary, as in case-sensitive keywords, for example.
- Do not capitalize the spelled-out form of an acronym unless specified otherwise in the **List of Acronyms and Abbreviations** (Appendix A).
- Avoid over-capitalization. The current practice is toward using lowercase unless there's a specific reason for capitalizing.

After consulting your project style sheet and this guide, use *American Heritage Dictionary* as the primary reference for proper capitalization of specific words and *The Chicago Manual of Style* for general guidelines.

Capitalization of interface elements

The following general guidelines cover the basic capitalization rules as they apply to interface elements:

- Menu names, command and command button names, and dialog box titles and tab names: Follow the interface. Usually, these items use title caps. If the interface is inconsistent, use title caps.
- Dialog box elements: Follow the interface. Newer style calls for these items to use sentence caps. If the interface is inconsistent, use sentence caps.
- Functional elements: Capitalize the names of functional elements that do not have a label in the interface, such as toolbars (the Standard toolbar) and toolbar buttons (the Insert Table button). Do not capitalize interface elements used generically, such as *toolbar, menu, scroll bar,* and *icon.*
- User input and program output: Do not capitalize unless it is case-sensitive.

Always consult your project style sheet for terms that may be case-sensitive or traditionally all uppercase or lowercase.

SEE ALSO **Dialog Boxes and Property Sheets, Document Conventions, Menus and Commands**

Capitalization of titles and headings

Many books and Help topics now capitalize only the first word of chapter titles and other headings; design guidelines are less formal than in the past. The following guidelines represent traditional title capitalization standards. They are especially useful in answering questions about capitalization of adverbs, prepositions, verbal phrases, and the like. If your design does not use traditional capitalization, follow your design guidelines.

- Capitalize all nouns, verbs (including *is* and other forms of *be),* adverbs (including *than* and *when),* adjectives (including *this* and *that),* and pronouns (including *its).*
- Always capitalize the first and last words, regardless of their part of speech ("The Text to Look For").
- Capitalize prepositions that are part of a verb phrase ("Backing Up Your Disk").
- Do not capitalize articles *(a, an, the)* unless an article is the first word in the title.
- Do not capitalize coordinate conjunctions *(and, but, for, nor, or).*
- Do not capitalize prepositions of four or fewer letters.
- Do not capitalize *to* in an infinitive phrase ("How to Format Your Hard Disk").
- Capitalize the second word in compound words if it is a noun or proper adjective or the words have equal weight *(Cross-Reference, Pre-Microsoft Software, Read/Write Access, Run-Time).* Do not capitalize the second word if it is another part of speech or a participle modifying the first word *(How-to, Take-off).*
- Capitalize interface and program terms that ordinarily would not be capitalized, unless they are case-sensitive ("The fdisk Command"). Follow the traditional use of keywords and other special

terms in programming languages ("The printf Function," "Using the EVEN and ALIGN Directives").

- In table column headings, capitalize only the first word of each column heading.

SEE ALSO **Lists, Tables**

Capitalization and punctuation

Do not capitalize the word following a colon unless the word is a proper noun or the text following the colon is a complete sentence.

Do not capitalize the word following an em dash unless it is a proper noun, even if the text following the em dash is a complete sentence.

Always capitalize the first word of a new sentence following any end punctuation. Write sentences to avoid the use of a case-sensitive lowercase word at the beginning.

Correct

The printf function is the most frequently used C function.

Incorrect

printf is the most frequently used C function.

card

Use in reference to *sound card* or *network card.* Do not use *board,* which can be ambiguous.

carriage return/line feed (n)

Follow conventional practice and use a slash mark, not a hyphen, when referring to this ASCII character combination. Use the acronym CR/LF for subsequent references.

carry out vs. run

For commands, use *carry out.* For programs and macros, use *run.* Avoid using *execute* for these operations, especially in end-user documentation. *Execute* is acceptable in programmer documents.

cascade (v)

Use sparingly except to refer to the Cascade command or to describe cascading style sheets.

SEE ALSO **cascading style sheets**

cascading menu

Avoid. Refers to a submenu, which is a menu that "cascades" from another menu. Use *submenu* instead.

SEE ALSO **Menus and Commands, submenu**

A B C D E F G H I J K L M N O P Q R S T U V W X Y Z

cascading style sheets

A method authors and readers can use to attach styles such as specific fonts, colors, and spacing to HTML documents. Because they "cascade," some elements take precedence over others.

Cascading style sheets (note that the phrase is lowercase) is an accepted industry term. The file extension for the style sheets is .css, but do not use *CSS* as an abbreviation to refer to the style sheets.

catalog

Not *catalogue.*

category axis

In spreadsheet programs, refers to the (usually) horizontal axis in charts and graphs that shows the categories being measured or compared. For clarity, refer to it as the "category (x) axis" at first mention; "x axis" is acceptable for subsequent mentions. You can also use "horizontal (x) axis" in documentation for novices.

SEE ALSO **value axis**

caution

Advises users that failure to take or avoid a specified action could result in loss of data.

In online documentation and messages, precede a caution with the warning symbol. See **Notes and Tips**.

Warning symbol

CBT

Means *computer-based training.* Avoid; use *tutorial* instead. Use *online tutorial* only to distinguish from a printed tutorial.

CD Plus

Refers to a method of combining audio and data on one compact disc. In general, use the term as an adjective, as in "CD Plus format," "CD Plus technology," and so on. Note spelling—not *CD+, CD-plus,* or other variations.

CD-ROM, CD-ROM drive

CD-ROM is the acronym for *compact disc read-only memory.* When referring to the disc itself, use *CD-ROM, compact disc,* or *disc* (not *disk*). Do not use the redundant *CD-ROM disc.*

Refer to the drive for a compact disc as the *CD-ROM drive,* not *CD-ROM player* or *CD drive.*

It is acceptable to use the abbreviation *CD* alone if there's no possibility of confusion with other compact discs, such as *CD-R* (recordable).

The components of some Microsoft CD-ROM programs (for example, Bookshelf and Programmer's Bookshelf) are *tools,* as in "tools for writers" and "tools for programmers."

cellular phone

Use *cellular phone,* not *cell phone, digital phone, mobile phone,* or *cellular telephone.*

center around

Do not use; use *center on* instead.

certificate

A digital certificate binds a client's or server's identity to a pair of electronic keys that can be used to encrypt and sign digital information. Certificates ensure secure, tamper-proof communication on the Internet. A certificate is obtained by a process called "code signing."

The certificate identifies the author and software publisher, so the user can contact them if the code is not satisfactory.

For more information about security, see the Microsoft Security Advisor at //www.microsoft.com/security.

chapter

Use only in reference to printed documents. For online documents, use *section, topic, site,* or other appropriate term.

channel

Use lowercase to refer to the channels on MSN or Internet Explorer 4.0 ("the Arts & Entertainment channel"). Use uppercase when necessary to match the interface ("the Channel bar").

character set

Do not use as a synonym for *code page.* A character set appears on a code page. For more information, see the *Microsoft Press Computer Dictionary.*

chart (n)

Do not use as a verb when referring to entering data for a graphic; use *plot* instead. Use the noun *chart* instead of *graph* to refer to graphic representations of data—for example, *bar chart, pie chart,* and *scatter chart.*

chat

"To chat" is to "talk" in real time by typing messages into the computer that others can see and respond to. A "chat room" is a service offered by many Internet service providers. Many chat rooms are dedicated to specific topics of conversation.

It's okay to use these terms in this context.

check

Do not use as a verb when referring to a check box in a dialog box; use *select* or *clear* instead. Do not use as a noun to mean *check mark.* Okay to use as an adjective, as in "checked commands."

SEE ALSO **checked command**

check box

Select and *clear* check boxes (not *turn on* and *turn off, mark* and *unmark, check* and *uncheck,* or *select* and *deselect*). Also, use the identifier *check box,* not just *box,* to refer to this option, because *box* alone is ambiguous for localizers.

See **Dialog Boxes and Property Sheets**.

Check boxes

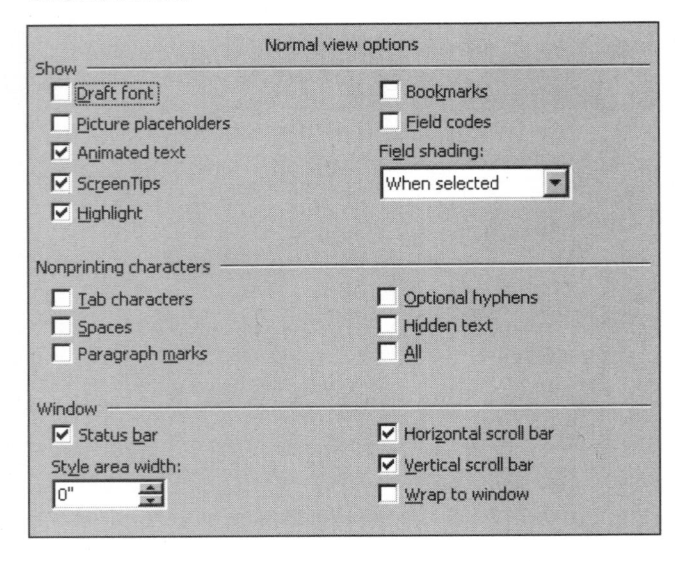

check mark

Two words.

checked command

A command name preceded by a check mark or a bullet that turns on or off each time the user chooses that command. Checked commands can be either mutually exclusive (document views in Word, for example) or independent of each other (settings on the View menu in Microsoft Excel). In the latter case, it's okay to call them *marked commands.*

Use "turn on" or "turn off" in procedures to refer to activating or deactivating the command, but "click" as the means of turning it on or off.

Correct

To turn on Outline view
On the View menu, click Outline.

On the View menu, click Outline.
A bullet means that the document is in Outline view.

child folder

Avoid. Use *subfolder* or *subdirectory* instead.

choose

In general, use *click* or *double-click* instead of *choose* for commands and buttons that carry out commands; compare with **select**.

SEE ALSO **click, Dialog Boxes and Property Sheets, point**

clear

Use *clear* for check boxes; do not use *turn off, unmark, uncheck,* or *deselect.*

Correct

Click to clear the Mirror margins check box.
To clear a tab stop, click Clear.

SEE ALSO **Dialog Boxes and Property Sheets**

click

In general, use *click,* rather than *choose* or *select,* to refer to a user's choosing or selecting a command or option.

If a user can set an option to use either a single click or a double-click, use the default mode when documenting a feature. But do explain the various options in Help and in the user's guide.

Do not use *click on* or *click at;* "click in the window" is acceptable, however.

> **NOTE** It's okay to omit "click OK" at the end of a procedure if the interface makes it clear that clicking the button is necessary to complete the procedure.

SEE ALSO **Dialog Boxes and Property Sheets, point, tap**

A B C D E F G H I J K L M N O P Q R S T U V W X Y Z

clickstream

One word. Refers to the path users take when browsing the Internet. Each click adds to the stream.

Avoid in end-user documentation.

client

Use *client* to refer only to a computer or to a client object, application, and such, not a person. Use a term such as *client workstation* or *client computer* if *client* alone is ambiguous.

Use *customer* to refer to a person.

client area

Do not use unless necessary and then in technical documentation only; use *desktop* or *workspace* instead.

client/server

Use the slash mark in all instances.

client side (n), client-side (adj)

Avoid if possible, especially as an adjective. Use *client* instead.

It's acceptable when a term has a specific technical meaning, as in "client-side image map."

clip art

Two words.

Clipboard

Capitalize when referring to the program, in both Windows and Macintosh documentation. Do not precede with *Windows.* Material is moved *to* the Clipboard, not *onto* it.

close

Use *close* for windows, documents, and dialog boxes. For programs, use *quit.* For network connections, use *end.*

close box

Macintosh only. A square box at the upper-left corner of a document window. Note lowercase.

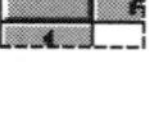

Close button

In Windows 98–based programs, the box with an *X* at the upper right of the screen. Okay to only show the Close button, without naming it, in procedures. However, if you spell out the command in reference to the button, use the word *button* because the button itself is unnamed.

Correct

... and then click [X].
... and then click the Close button.

co (prefix)

In general, do not hyphenate words beginning with *co,* such as *coauthor* and *coordinate,* unless it's necessary to avoid confusion. If in doubt, check *American Heritage Dictionary* or your project style sheet.

Code Commenting Conventions

Many code commenting conventions are dictated by practices in particular languages. Instead of prescribing commenting conventions across groups and languages, these guidelines call attention to issues each group should consider as it develops its own commenting conventions.

> **NOTE** It is acceptable to say that some code has been "commented out," but do not use "REM'd out." A REM (remark) statement is specific only to a few languages.

General rules

Use a consistent commenting style within and across your development product's documentation components.

Decide what the general function of your code comments should be. In software development, developers should use comments to document things that aren't apparent from reading the code itself—for example, the algorithm being used, the rationale for a routine, the meaning of a flag set to 0, or the content of a temporary variable. In this setting, avoid merely paraphrasing the code in comments.

In documentation, however, comments may need to be more explicit and more educational—documenting what a routine does, for instance.

Comment that provides a rationale not apparent from the code itself

```
' Remove the name from the collection. Because collections are reindexed
' automatically, remove the first member on each iteration.
For Num = 1 To MyClasses.Count
    MyClasses.Remove 1
Next
```

A
B
C
D
E
F
G
H
I
J
K
L
M
N
O
P
Q
R
S
T
U
V
W
X
Y
Z

Comments that paraphrase the code

```
' Function procedure definition.
Function ReturnTwice( Optional A )
        If IsMissing( A ) Then
                ' If the argument is missing, return a Null.
                ReturnTwice = Null
        Else
                ' If the argument is present, return twice the value.
                ReturnTwice = A * 2
        End If
End Function
```

Decide whether you'll use endline comments and to what extent. Developers and documentation writers in many programming languages (except assembly language) try to use a minimum of endline comments, which are most useful for describing variables in data declarations and for explaining techniques used in implementations. Thus, developers and writers often rely on paragraph comments if that's practical or provide explanation in accompanying text.

Paragraph comments

If you decide to use paragraph comments, you'll also have to decide:

- Whether to use full sentences with standard English capitalization and punctuation for program, file, module, and routine paragraph comments.
- Whether to put a space between the comment delimiter and the beginning of the comment text.
- Whether to align a paragraph comment with the code it comments on.
- Whether to use line spaces above and below (and, if so, how many) the paragraph comments at the various levels of your code—program, file, module, routine, block, and line.

Endline comments

If you decide to use endline comments, consider the following:

- Decide whether the comments should be sentence fragments or complete sentences.
- Strive for parallel construction: Try to use all verb phrases or all noun phrases, for example, or all declarative constructions or all imperative constructions.
- Decide whether to use sentence caps and end punctuation.
- Decide whether to omit articles *(a, and, the)*, and, if so, omit them consistently.
- Decide whether to use abbreviations; if yes, develop a consistent abbreviation style and use the abbreviations consistently.
- Decide whether to left align endline comments.
- Decide whether to put a space between the comment delimiter and the comment text. This decision will usually be congruent with your practice in paragraph comments.
- Decide whether to allow multiline endline comments or to convert them to paragraph comments. If you do decide to use a multiline endline comment, postpone the next line of code until the

multiline endline comment is complete. (Note that postponing the next line of code can force a linespace at odds with the logical structure of the code.) Use capitalization, punctuation, or spacing or some combination of the three to distinguish multiline endline comments from successive single-line endline comments.

Multiline endline comments

```
LeftEdgeDir = 1;        // Left edge runs up through vertex list
Temp = MinIndexL

MinIndexL = MinIndexR;     // Swap indices so that MinIndexL
MinIndexR = Temp;          // points to start of left
                           // edge; same for MinIndexR
.
.
.
m_pVBGrid->SetStrProperty( "Text", temp );  // Copy new value
                                            // to selected cell
ComputeSums();
.
.
.
```

Multiline comments converted to paragraph comments

```
// The left edge runs up through the vertex list.
LeftEdgeDir = 1;
Temp = MinIndexL;

// Swap the indices so that MinIndexL points to the start
// of the left edge; the same for MinIndexR.
MinIndexL = MinIndexR;
MinIndexR = Temp;
.
.
.
// Copy the new value to the selected cell.
m_pVBGrid->SetStrProperty( "Text", temp );
ComputeSums();
```

SEE ALSO **Code Formatting Conventions, Coding Style**

For More Information

McConnell, Steve. *Code Complete.* Redmond, WA: Microsoft Press, 1993.

Code Formatting Conventions

In development tools documentation, programming elements must stand out from the text. Most programming elements are formatted according to document conventions: constants, variables, functions, statements, type names, directives, instructions, operators, macros, format specifiers, and any other predefined or user-defined element.

Sometimes, however, document conventions conflict. For example, you might want to refer to something the user must type, in a generic sense, such as "file name." Is that *filename,* **filename**, `file_name`, or some combination thereof? One additional convention for programming elements is monospace text, which can appear within or set off from normal text.

Here are some cases of conflicting conventions; consider how you will treat them in your documentation.

Formatting monospace text

If programming elements consist of less than a logical line, you can format them within normal text as monospace (Lucida Sans Typewriter). Use a point size that works with your normal text. Format the spaces on either side in the same font as the normal text.

If you choose monospace text, continue to set complete lines or statements of sample code on their own; do not embed them in normal text. Complete the sentence that introduces the sample code.

Sample code

Sample code provided for the user is in monospace text. Continue to use monospace text to refer to a portion of the sample code, if the portion is less than a logical line or a full command. If the portion you want to draw attention to can stand on its own as a line or statement, place it on its own line. To show portions of sample code, use horizontal or vertical ellipses.

Correct

In this example, a form named `myForm` contains a command button named `Command1`. The following command changes the Top property of `Command1` to 50.

```
=ChangeTop(myForm.Command1, 50)
```

Keywords in text

For keywords within text, follow document conventions. For keywords in code samples, follow coding conventions.

Correct

This code tries to create a char pointer named `ptr` and initialize it with a string:

```
main()
{
    char *ptr;
    strcpy( ptr, "Ashby" );  /* Error! */
}
```

However, the declaration `char *ptr` creates a pointer but does nothing else.

Syntax

Syntax itself follows document conventions. To refer to a part of the syntax, echo the document conventions within the line of normal text.

Correct

int _SetObjectProperty(Value FAR **object,* char FAR **prop,* Value FAR **val,* int *fAdd*);

If *fAdd* is a nonzero value and the property you specify does not exist for the object, then the property is added to the object as a user-defined property and contains the property value you specify with **val.*

User output

If code typed at the command line or in a command window produces output, consider how you will distinguish code the user types from text that is displayed as a result. The following example, from Visual FoxPro, uses monospace text for both the input and output and a heading to separate the two sections.

Correct

Because `cFirst` is character data and `nSecond` is numeric data, you get a data type mismatch error if you type the following command:

```
? cFirst + nSecond
```

You can avoid this problem by using conversion functions. These functions enable you to use the following operations:

```
? cFirst + LTRIM(STR(nSecond))
? VAL (cFirst) + nSecond
```

Output

```
12345
168
```

Annotated code

If you want to show the user how to alter an existing code sample, consider how to demonstrate which lines have changed. The following example, from Visual C++, uses arrow-shaped dingbats to indicate the significant lines of code.

Correct

Add OnDraw to file Scribvw.cpp, as defined here:

```
void CScribView::OnDraw( CDC* pDC )
{
        CScribDoc* pDoc = GetDocument();
        ASSERT_VALID(pDoc)
⇒       // The view delegates the drawing of individual strokes to
```

(continued)

```
⇒       // CStroke::DrawStroke( ).
⇒       CTypedPtrList<CObList, CStroke*>& strokeList = pDoc->m_strokeList;
⇒       POSITION pos = strokeList.GetHeadPosition( );
⇒       while (pos != NULL)
⇒      {
⇒         CStroke* pStroke = strokeList.GetNext(pos);
⇒         pStroke->DrawStroke( pDC );
⇒       }
}
```

SEE ALSO **Code Commenting Conventions, Coding Style, Command Syntax, Document Conventions**

code page

Do not use as a synonym for *character set.* A character set appears on a code page. For more information, see *Microsoft Press Computer Dictionary.*

code signing

Technical term. Method of obtaining a digital certificate for software that ensures it is not tampered with while downloading from the Internet.

Coding Style

Most development tools groups have already developed coding style conventions appropriate to their languages. Acknowledging the inadvisability of recommending a single coding style across groups and languages, these guidelines call attention to the issues each group should consider as it develops conventions for the code listings in the documentation for its product.

General rules

Apply a consistent coding style within a document and across documentation for your development product.

In addition to observing the syntax requirements of your language, use the coding style of the browser that will be used with your product to guide your coding style choices.

Indentation and alignment

Consistent indentation and alignment can make your code easier to read. These are some guidelines to consider for indentation and alignment:

- Use indentation consistently.
- Develop a scheme for positioning the opening and closing braces of a block of code.
- Develop conventions for handling statements that wrap to the next line. You may decide, for instance, to break and indent the line so that elements are aligned for logical parallelism rather than according to the indentation scheme.

- Decide whether you'll declare one variable per line even if the variables are of the same type.
- Decide whether variables of the same type that are logically related can be declared on the same line.
- Decide whether you'll permit If-Then statements to appear on the same line.
- Decide whether you'll put the keywords *else* and *do* on their own lines.

Line spacing

Develop line-spacing conventions that help users see the hierarchical relationships in your code and help distinguish between declarations and implementation.

Spacing within lines

These are some places where you'll have to decide whether to use spaces within lines:

- Before and after an operator.
- Before and after an array subscript.
- After the commas in an array that has multiple dimensions.
- Between a function name and an initial parenthesis that follows it.
- Between a language keyword and a parenthetical expression that follows it.
- After the initial parenthesis and before the final parenthesis of required parentheses.
- After the initial parenthesis and before the final parenthesis of parentheses that specify or clarify evaluation order ("optional" parentheses).

Naming and capitalization

Establish consistent naming and capitalization schemes for variables, constants, user-defined data types, classes, functions, and user-defined functions. Avoid confusing abbreviations such as *No,* which can mean either "no" or "number."

Also establish consistent naming and capitalization schemes for special programming elements such as character constants, string constants, prefixes, and suffixes.

Finally, establish conventions for octal and hexadecimal notation.

SEE ALSO **Code Commenting Conventions, Code Formatting Conventions**

For More Information

McConnell, Steve. *Code Complete.* Redmond, WA: Microsoft Press, 1993.

collaborate (v), collaboration (n)

Use when referring to two or more people working in a document over the Internet.

Colons

Use a colon at the end of a sentence or phrase that introduces a list. Do not use a colon to introduce art, tables, or sections or following a procedure heading.

Within text, use colons sparingly: A colon in text usually signifies that what follows illuminates or expands on that statement.

Correct

The basic configuration for your computer system includes:

- A hard disk with 24 megabytes of free disk space and at least one floppy disk drive.
- A monitor supported by Microsoft Windows.

Use a colon at the end of a sentence or phrase introducing sample code, unless other text intervenes.

Correct

For example, to open the external FoxPro database on the network share \\FoxPro\Data in the directory \Ap, use the following:

```
... code sample
```

Capitalization after a colon

Do not capitalize the word following a colon within a sentence unless the word is a proper noun or the text following the colon is a complete sentence, as in the second paragraph. Do capitalize the first word of each item in a list, however.

Use only one space after a colon in both online and printed documentation.

color map

Two words. Refers to the color look-up table in a video adapter.

column format

Not *columnar* or *columnlike.*

COM

As a device name, use all uppercase followed by a number, as in *COM1.* As an extension and the indicator of a commercial organization in a URL, use all lowercase preceded with a period, as in *.com file* and *microsoft.com.*

COM is also an acceptable acronym for *Component Object Model.* See **COM, ActiveX, and OLE Terminology**.

COM, ActiveX, and OLE Terminology

The Component Object Model (COM) is a collection of services that let software components interoperate in a networked environment. COM-based technologies include distributed COM, ActiveX Controls, Microsoft Transaction Server (MTS), and others. See //www.microsoft.com/com for more information about COM.

ActiveX is a way to package components for delivery in Web browsers or other software using COM. OLE, which is also built on COM, is a subset of ActiveX. It is used primarily for linking and embedding to create compound documents for easier use and integration of desktop programs.

COM terminology constantly evolves, but in general many COM terms have replaced ActiveX terms, which in turn replaced most OLE terms. Some terms specific to ActiveX and OLE do remain in use, however.

NOTE This terminology relates in precise ways to the technologies concerned. It should not be used in end-user documentation.

General guidelines

- Use *ActiveX* only in the context of ActiveX controls. Otherwise, use *Active*, unless this table lists an exception.
- ActiveX is a registered trademark. Use the ® symbol at first mention.
- Do not make changes that can break existing code (for example, of class names, properties, methods, or events). Do not change any third-party name that includes OLE or OCX.
- Do not use any OLE-specific terms other than those listed in this table.
- Follow the capitalization in the tables for references to COM and ActiveX technologies.

Term	*Definition*	*Usage (and Queries)*
active content	Synonym for active objects, active scripts, and active documents.	
active document	A document that contains ActiveX controls, Java applets, HTML pages, or document objects. Currently, the only containers that can display active documents are Internet Explorer and Microsoft Binder.	Do not use *ActiveX document.*
active script	A script that can be implemented in various languages, syntaxes, persistent formats, and so on that can interact with other ActiveX controls	
active scripting	Microsoft technology that uses COM to connect third-party scripts to Microsoft Internet Explorer without regard to language and other elements of implementation.	Do not use *COM scripting, ActiveX scripting,* or *OLE scripting.*

(continued)

Term	*Definition*	*Usage (and Queries)*
Active Server Pages (.asp file)	An Internet Information Server (IIS) feature that combines HTML and active scripts or components.	Treat the term as a singular (but do not use *Active Server Page*). Use *.asp file* to refer to the file that generates the Web page and the acronym phrase *ASP page* to refer to the Web page that actually appears in the browser.
ActiveX-based, ActiveX-enabled		Do not use. Use COM-based or write around instead.
ActiveX component	See **COM component**, on the next page.	Do not use. Use *COM component* instead.
ActiveX control	User interface element created using ActiveX technology. The only time to use *ActiveX* is in the context of an ActiveX control.	Do not use *COM control*. There is no such thing. You can use *control* alone if the context is clear.
ActiveX Controls	Technology for creating UI elements for delivery on the Web.	Capitalize *Controls* when referring to the technology, but not when referring to specific controls.
ActiveX Data Objects (ADO)	High-level data access programming interface to the underlying OLE DB data access technology, implemented using COM.	*ActiveX* is correct in this instance. Usually referred to by the acronym, however.
Automation	COM-based technology that enables dynamic binding to COM objects at run time.	Do not use *ActiveX Automation* or *OLE Automation*
Automation controller	An application or programming tool that accesses Automation objects.	Obsolete term; do not use. Use *Automation client* instead.
Automation object	An object that is exposed to other applications or programming tools through Automation interfaces.	Use instead of *programmable object.*
client	Generic term referring to any program that accesses or uses a service provided by another component.	Do not use *client component;* instead use a phrase such as "is a client of" or "acts as a client." A program is a client in relation to some server object, not in a vacuum.
COM class	Definition in source code for creating objects that implement the IUnknown interface and expose only their interfaces to clients.	

Term	Definition	Usage (and Queries)
COM component	Binary file containing code for one or more class factories, COM classes, registry-entry mechanisms, loading code, and so on.	Do not use *ActiveX component.*
COM object	Instance of a COM class.	
component	Code module that serves up objects.	Do not use as a synonym for object.
component object		Do not use; use either *component* or *object* instead, whichever is accurate.
compound document (OLE-specific term)	A document that contains linked or embedded objects, as well as its own data.	Do not use as a synonym for compound file. (See **compound file**, on this page.)
compound document object (OLE-specific term)		Obsolete term; do not use. Instead, use *linked* or *embedded object.*
compound file (OLE)		Avoid; *compound file* is too easily confused with *compound document.* Compound document refers to the default implementation of structured storage. If necessary to refer to the concept, explain it before using the term.
container	An application or object that contains other objects, for example, a client of a linked or embedded object (synonym for *compound document*).	
custom control		Obsolete term. See **ActiveX control**, on page 50.
DCOM	Refers to "distributed COM," a wire protocol that enables software components to communicate directly over a network.	Use *DCOM* only to refer specifically to the wire protocol. Use just *COM* in all instances to refer to the technology for distributing such components.
docfile (OLE-specific term)	Obsolete synonym for *compound file.*	Do not use. See **compound file**, on this page.
dynamic COM component	COM object created by a run-time environment that is not explicitly defined as a COM class.	**NOTE** Term may change to *dynamic COM object.*
embed (OLE)	To insert an object in its native format into a compound document. Contrast with **link object**, on page 52.	

(continued)

Term	*Definition*	*Usage (and Queries)*
embedded object (OLE)	An object whose data is stored along with that of its container but that runs in the process space of its server.	
expose	To make an object's services available to clients.	
in-place activation	Also called *visual editing*. The process of editing an embedded object within the window of its container, using tools provided by its server. Note that linked objects do not support in-place activation.	Do not use *in-situ editing*. You can also describe the process, for example, "editing embedded objects in place."
in-process component	A component that runs in its client's process space.	Do not refer to specific file types as *in-process*. For example, don't use *in-process DLLs*. Instead, use *in-process component*. *In-process server* is okay.
insertable object (OLE)		Avoid except to match the interface in end-user documentation.
invisible object		Avoid; instead, use a phrase such as "an object without a visible user interface."
link object (OLE)	An object that specifies and maintains the relationship between a **linked object** and a **link source** (see entries on this page).	Avoid using link as a verb in OLE contexts. Instead, use a phrase such as "create a link to."
link source (OLE)	A data object stored in a location separate from the container and whose data is represented in the container by a linked object. When the data in the link source is updated, the corresponding data representation in the linked object can be updated automatically or manually.	
linked object (OLE)	An object that represents a link source's data in a compound document. The link source is identified by an associated link object.	Differentiate between *link object* and *linked object*.

Term	Definition	Usage (and Queries)
local component		Do not use. Instead, use *out-of-process component* or *out-of-process server.* Refer to machine location only if it's important for a particular discussion. Use *in-process component* to refer to a component that runs in the same processing space as its client.
member function		Avoid; use *method* instead.
object	A combination of code and data created at runtime that can be treated as a unit. An instance of a class.	Do not use as a synonym for *component.*
Object linking and embedding (OLE2.0 /Compound Documents)		Use only in this OLE-specific context.
object linking and embedding (OLE)		Do not use as a synonym for the OLE technology. Use sparingly and only when referring to the process itself.
object model		Do not use; instead, use *Component Object Model (COM). Object model* is a general term; *COM* is a specific object model.
OCX		Do not use as a synonym for an ActiveX control. Use of .ocx as a file extension is acceptable
OLE control, OLE Controls		Obsolete terms. In general usage, use *ActiveX control* or *control.* For the technology, use ActiveX Control. See **ActiveX control, ActiveX Controls**, on page 50.
OLE drag and drop	The technology for dragging and dropping COM objects.	Okay to use.
OLE object	An object that supports object linking and embedding and that can be linked and/or embedded.	Use only when referring to objects that are linked or embedded.
out-of-process component	A component compiled as an executable file to provide services.	Do not use EXE server.

(continued)

A B C D E F G H I J K L M N O P Q R S T U V W X Y Z

Term	Definition	Usage (and Queries)
remote component		Avoid; use *out-of-process component on a remote machine.* But use that only when location is important, which is seldom.
server	A synonym for *component*. The term *server* emphasizes a relationship to a client so is useful when this relationship is important.	Do not use without a modifier. In general, avoid *COM server* except in some instances in programmer documentation. Use *COM object* or *COM component* instead. If you use *COM server,* make sure that your usage will not be confused with a server computer.
server application	See **server**, on this page.	Avoid; instead, use *COM object, COM component,* or just *component,* unless you are specifically comparing to a client application. Obsolete term from the early days of OLE.
visual editing	See **in-place activation**, on page 52.	

combo box

Technical term for the dialog box option that's a text box with an attached list box. Do not use the term in end-user documentation. Use the name of the box with the word *box* instead, as in "the Font box."

Combo box

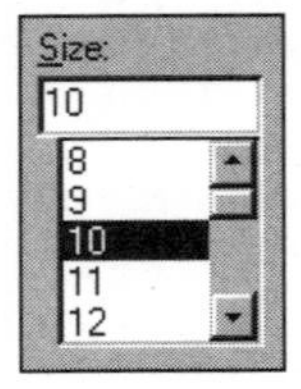

SEE ALSO **Dialog Boxes and Property Sheets**

command

Use *command*, not *menu item, choice,* or *option.* Users *click* commands, or if you're documenting both mouse and keyboard instructions, they *choose* commands. See **Menus and Commands**.

command button

Refers to the usually rectangular button in a dialog box that carries out a command. Generally, just use the command name, without the word *button,* especially in procedures. Follow the interface for capitalization and spelling.

Don't refer to a command button as an *action button* or *push button.* In technical documentation, it's acceptable to say something like "a command button is also called a push button." See **Dialog Boxes and Property Sheets**.

Command button

command prompt

Not *C prompt, command-line prompt,* or *system prompt.* If necessary, use a specific reference, such as "MS-DOS prompt."

Command Syntax

Syntax is the order in which the user must type a command-line command or utility name and any arguments and options that follow it. The user must type elements that appear in bold in the syntax line exactly as they appear. Elements that appear in italic are placeholders representing information the user must supply.

> **NOTE** In text-mode documents, where formatting is unavailable, substitute all uppercase for bold and all lowercase for italic.

This is a sample syntax line using the form standard in Microsoft documentation:

sample {**+r** | **–r**} *arguments* ... [*options*]

The meaning of each of these elements is as follows.

Element	*Meaning*
sample	Specifies the name of the command or utility.
{ }	Indicates a set of choices from which the user must choose one.
|	Separates two mutually exclusive choices in a syntax line. The user types one of these choices, not the symbol.
arguments	Specifies a variable name or other information the user must provide—for example, a path and file name.
...	Indicates that an argument can be repeated several times in a command line. The user types only the information, not the ellipsis (...).
[ ]	Indicates optional items, except in languages in which brackets are part of the syntax. In that case, use double brackets ([[ ]]). The user types only the information within the brackets, not the brackets.

A B C D E F G H I J K L M N O P Q R S T U V W X Y Z

Here are syntax lines for a Visual Basic property (FontSize) and a C library routine (_setfont).

Correct

{[*form.*] [*control.*]|**Printer.**}**FontSize**[=*points%*]

short far _setfont(unsigned char far ***options**);

Follow exactly the standards and conventions of the language or program that you are documenting when writing syntax.

Line breaks in syntax

Do not hyphenate a line of command syntax. If you must break a line, break it at a character space and do not use a hyphen. Indent the runover line.

Correct

Set *database* = **OpenDataBase**(*dbname*[, *reserved*[, *read-only*[,
connect]]])

Incorrect

Set *database* = **OpenDataBase**(dbname[, reserved[, read-
only[, connect]]])

Format of syntax

In general, follow these document conventions for formatting syntax.

For these elements	*Use*
Keywords, functions, and anything else that must be entered exactly as shown	Bold
Variables and other placeholders the user must provide	Italic
Punctuation marks the user does not enter	Roman

SEE ALSO **Code Formatting Conventions, Document Conventions, Procedures**

Commas

Comma usage is governed by both convention and grammar. For more details about comma usage, see *Harbrace College Handbook.*

When to use commas

In a series consisting of three or more elements, separate the elements with commas. When a conjunction joins the last two elements in a series, use a comma before the conjunction.

Correct

Chapter 15 is an alphabetical reference to commands, procedures, and related topics.
You need a hard disk, an EGA or VGA monitor, and a mouse.

Use a comma following an introductory phrase.

Correct

In Microsoft Windows, you can run many programs.

When not to use commas

Do not join independent clauses with a comma unless you include a conjunction. Online documentation often has space constraints, and it may be difficult to fit in the coordinate conjunction after the comma. In these instances, separate into two sentences or use a semicolon.

Correct

Click Options, and then click Allow Fast Saves.
Click Options; then click Allow Fast Saves. [only to save space in online documentation]

Incorrect

Click Options, then click Allow Fast Saves.

Do not use a comma between the verbs in a compound predicate.

Correct

The Setup program evaluates your computer system and then copies the essential files to your hard disk.
The Setup program evaluates your computer system, and then it copies the essential files to your hard disk.

Incorrect

The Setup program evaluates your computer system, and then copies the essential files to your hard disk.

Do not use commas in month-year formats.

Correct

Microsoft introduced Microsoft Windows version 3.0 in May 1990.

Incorrect

Microsoft introduced Microsoft Windows version 3.0 in May, 1990.

communications port

Acceptable to abbreviate as *COM port* after first mention. Be sure the context is clear so that there's no confusion with *COM* meaning "component object model."

compact disc

Not *CD disc* or *CD-ROM disc*. It's acceptable to use *CD* if there's no possibility of confusion with other disc formats.

SEE ALSO **CD-ROM, CD-ROM drive**

A B C D E F G H I J K L M N O P Q R S T U V W X Y Z

compare to vs. compare with

Use *compare to* to point out similarities between unlike items. Use *compare with* to comment on the similarities or differences between similar items. The use of *compare to,* which is often metaphorical, is generally unnecessary in documentation.

Correct

People have compared a computer to a human brain.
Compared with a Pentium 90, a 386 is extremely slow.

compile (adj, v)

Do not use as a noun, as in "do a compile." Acceptable to use as an adjective, as in "compile time."

comprise

Avoid in general, mostly because its meaning is often confused. It means "include" or "contain." Depending on your meaning, use those terms or *consist of* or *make up* as appropriate. Do not use *comprised of.*

computer

Not *PC* or *machine* or *box.*

context menu

Do not use to refer to the menu that appears when a user clicks the right mouse button in certain areas (or "contexts"), such as in a toolbar region. Use *shortcut menu* instead. Do not use *pop-up menu.*

context-sensitive (adj)

Hyphenate in all positions—that is, both as an adjective and as a predicate adjective.

contiguous selection

Avoid; this term may be unfamiliar to many users. Use *adjacent selection* instead.

Contractions

Use of contractions is at the discretion of the editor. Advantages of contractions are that they lend an informal tone and save space. However, readability experts say that a reader requires more time to decode a contraction than a complete phrase.

Follow these rules for using contractions:

- Always use an apostrophe to indicate where letters have been deleted to form the contraction—for example, *can't* (from *cannot*) and *it's* (from *it is*).
- Never use a contraction with *Microsoft* or with a product name, as in "Microsoft's one of the fastest-growing companies in the computer industry" or "Microsoft Excel's the best-selling spreadsheet program."
- Never form a contraction from a noun and a verb. That is, never use a construction such as "Microsoft'll release a new version soon" or "the company's developing a lot of new products."
- You can, however, form contractions from pronouns and verbs, as in "If you're in Design view" or "If it's pressed in."

control

In technical documentation, the generic term for most dialog box elements, such as options. Do not use with this meaning in end-user documentation.

In some hardware products, particularly joysticks, *control* is correct to refer to buttons, switches, and other elements with which the user "controls" various actions, especially moving around the screen.

control-menu box

Used to describe the button at the far left on the title bar in Windows 3.1–based programs. In Windows 95–based programs, this box has been replaced by the program icon, which is used primarily when working with multiple open windows. Avoid referring to this icon if possible.

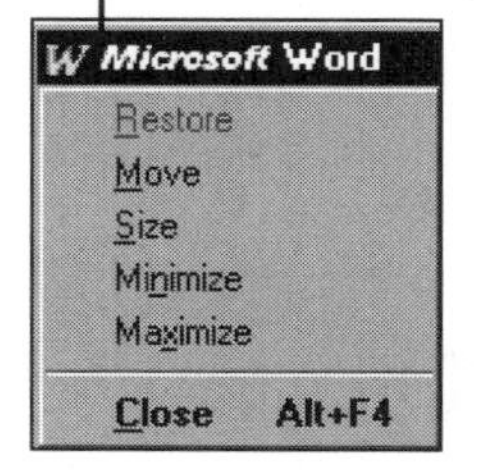

SEE ALSO **Screen Terminology**

Control Panel

Do not use *the* when referring to Control Panel.

Control Panel (Web view)

Correct

Click Control Panel, then double-click the icon whose settings you want to change.

Note that there is only one Control Panel in Windows. The icons within it are not separate "control panels."

Correct

In Control Panel, click the Network icon .

Incorrect

Click the Network Control Panel.

Copyright Pages and Copyright Screens

Copyright protection is granted to any original work of authorship fixed in any tangible medium of expression from which it can be perceived, reproduced, or communicated. Although legally copyright is assumed and a copyright notice is not required to provide protection, Microsoft continues to include copyright notices on all of its works.

Copyright notice for online and printed documentation

The copyright notice states that the text or program code contained within the item on which the notice appears is the intellectual property of Microsoft. The copyright notice usually takes the following form:

© 1985–1998 Microsoft Corporation. All rights reserved.

> **NOTE** Only the copyright symbol ©, not the word *Copyright,* is required in copyright notices. Although some groups also include the word *Copyright,* doing so does not afford any protection in addition to the symbol alone and is therefore unnecessary.

If a product contains audio, accompany the © with a ℗ :

© & ℗ 1985-1998 Microsoft Corporation. All rights reserved.

Online, the copyright notice appears on the splash screen when a program is started and in the About dialog box on the Help menu of each program. Because online Help is copyrighted as part of the software product, its copyright notice should use the copyright date for the software itself.

Sample: Online copyright screen

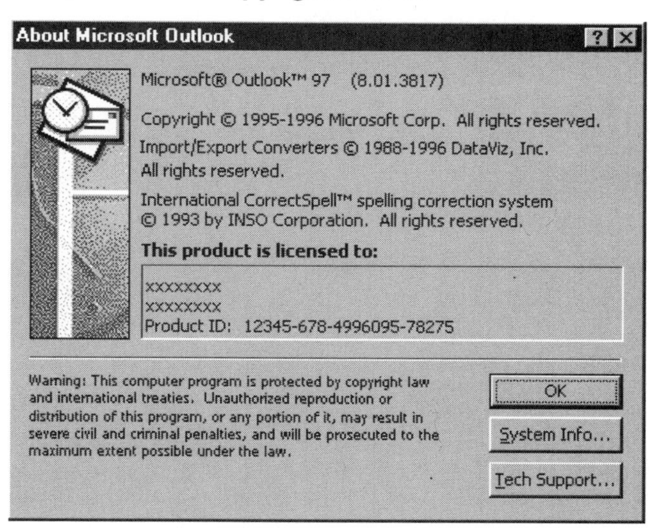

Printed documentation includes a copyright notice on a separate page, which is usually on the back of the title page or a similar location in front matter. CD-ROM books that replace printed documentation also include copyright screens that follow the guidelines for printed documentation.

Sample: Copyright page for printed documentation

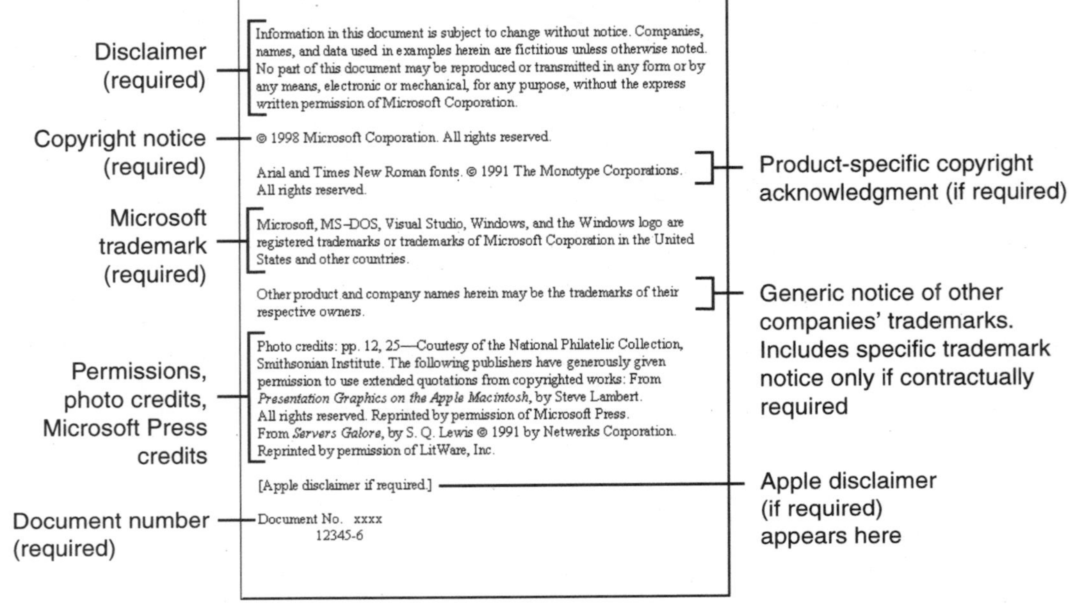

Information in this document is subject to change without notice. Companies, names, and data used in examples herein are fictitious unless otherwise noted. No part of this document may be reproduced or transmitted in any form or by any means, electronic or mechanical, for any purpose, without the express written permission of Microsoft Corporation.

© 1998 Microsoft Corporation. All rights reserved.

Arial and Times New Roman fonts. © 1991 The Monotype Corporations. All rights reserved.

Microsoft, MS-DOS, Visual Studio, Windows, and the Windows logo are registered trademarks or trademarks of Microsoft Corporation in the United States and other countries.

Other product and company names herein may be the trademarks of their respective owners.

Photo credits: pp. 12, 25—Courtesy of the National Philatelic Collection, Smithsonian Institute. The following publishers have generously given permission to use extended quotations from copyrighted works: From *Presentation Graphics on the Apple Macintosh*, by Steve Lambert. All rights reserved. Reprinted by permission of Microsoft Press. From *Servers Galore*, by S. Q. Lewis © 1991 by Netwerks Corporation. Reprinted by permission of LitWare, Inc.

[Apple disclaimer if required]

Document No. xxxx
12345-6

Copyright notice for World Wide Web pages

Each Microsoft Web page must include the standard Microsoft copyright notice at the bottom:

© 1998 Microsoft and/or its suppliers. All rights reserved. Terms of Use.

Determining the copyright date

For printed material, the original copyright date is the year the piece is first published. When reprinting material, follow these guidelines:

- If 85 percent or more of a piece is new at reprinting, it is considered a new work, and the copyright should list only the current year at the time of the reprinting. If less than 85 percent of a piece is new at reprinting, it is considered a derivative work, and the copyright should list both the original year of printing and the current year at the time of reprinting—for example, "© 1994, 1996 Microsoft Corporation."
- If material contains misinformation that could cause the software to fail or could cause a serious usability problem, the material should be reprinted immediately. If reprinting takes place in a year later than the one in which the piece was originally published, add the second year—for example, "© 1994–1995 Microsoft Corporation."
- If material contains a typographical or formatting error that is corrected in reprinting, keep the original copyright date.

corrupted (adj)

Do not use *corrupt* as an adjective.

Correct

The file may be corrupted.

crash

Do not use in end-user documentation; use *fail* for disks or *stop responding* for programs. In programmer documentation it may be the best word in certain circumstances, but it is computer jargon.

Acceptable

Although unaligned pointers degrade performance on 386 and 486 computers, they crash RISC-based computers.

criteria

Plural of *criterion*. It is acceptable to use *criteria* in database documentation to refer to one or more instructions about records.

Cross-References

Use cross-references in documentation to direct users to related information that may add to their understanding of a concept. However, try to write and edit so that you use cross-references only occasionally, for information that is not essential to the task at hand. For example, users should not have to look up material to complete a procedure. If the material has too many cross-references, restructure it. Technical material can have more cross-references than nontechnical material.

On the other hand, material on the World Wide Web may benefit from links to other Web pages and even to sections on a long page. Links used judiciously can shorten long pages and help users find exactly the material they want.

Different products have different needs and methods for referring to other material, so always consult your project style sheet for specific information about your group's cross-references.

Cross-reference syntax

Start a cross-reference by telling users why they should look elsewhere, not where they should look, even if the reason is not specific. For links to other Web sites, don't use a phrase such as "click here." Make the cross-reference text a hyperlink.

Correct

For more information about modifying Visual Basic source code and installing Visual Basic forms, see Chapter 9, "Extending Forms."
For more information, see Chapter 9, "Extending Forms."
For information about the accessibility appendixes, see the Accessibility intranet site.

Incorrect

See Chapter 9 for more information about extending forms.
See the Convert Notes dialog box for more information.
For information about the accessibility appendixes, click here.

Do not use *above* and *below* in cross-references in either printed or online documentation. Use *earlier* or *later,* give a more specific reference, or include a link to another Help topic or Web site.

Cross-reference forms for online documentation

Most online cross-references are jumps to other topics, such as to how-to information from overviews and conceptual information. In Windows 95–based programs, cross-references, usually titled something like "SEE ALSO," "Related Topics," or "Another Way," generally lead to the Topics Found window in the index. From there, the user can select a topic to go to. In other programs, "SEE ALSO" topics can appear either in a nonscrolling region after the topic title or at the end of the topic. This kind of reference can refer both to other Help topics and to book titles.

Topics Found window

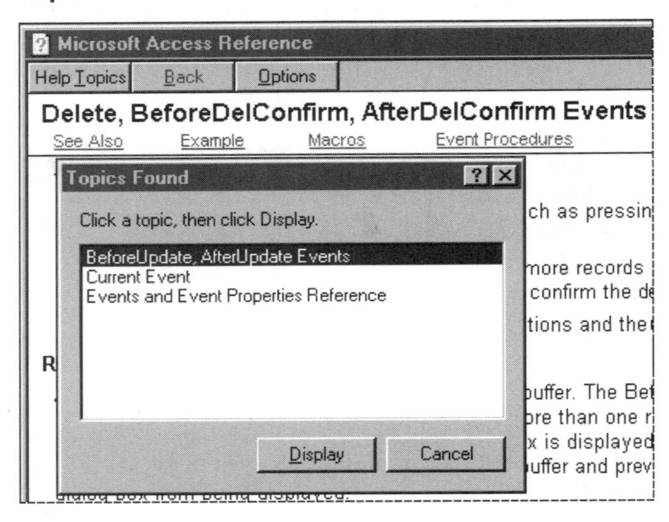

If you include a cross-reference jump or link within a topic, use wording similar to the following.

Correct

For more information, see Customizing and Optimizing Word.
For more information about adding fonts, see Customizing and Optimizing Word.

If the context is clear and if the link is color coded, a sentence such as "See Customizing and Optimizing Word" or "SEE ALSO OLE" is acceptable.

SEE ALSO **Bibliographies, Hyperlinks, Indexing, Keywords and Online Index Entries**

Cross-reference forms for printed documentation

There are a number of appropriate ways to handle cross-references in books:

- Within or at the end of a paragraph, when a cross-reference is appropriate to a specific context.

  Correct

  For information on arithmetic conversions, see "Usual Arithmetic Conversions," earlier in this chapter.

- In the margin, when a cross-reference applies to a topic or section that appears on that page.

  Correct

  SEE ALSO
  For more information about the Microsoft Download Service, see the "Accessibility for People with Disabilities" appendix.

- With a subheading, generally used in designs for technical references.

  Correct

  SEE ALSO **fgets, gets, puts**

- In a two-column table at the end of a chapter, when the cross-references are to more general related topics.

  Correct

To	*See*
Route a Word document for review	"Have Your Team Review a Word Document," page 333
Fax a copy of a document	"Create a Fax Cover Sheet," page 157
Add fractions, exponents, integrals, and other mathematical elements to a document	"Equation Editor" in the Word online index

If you are cross-referencing material within the same book and the Help file only, one table listing for both is enough. If you are cross-referencing a lot of material in many different documents, consider using a separate table for each document.

If you can, use page numbers in cross-references to specific sections or subsections, especially within the same book. Use the chapter number and title for more general information. If you use a page number, the chapter number and title aren't needed.

Correct

... see "Formatting an Outline," p. 226.

... see Chapter 17, "Outlining and Organizing a Document."

If your cross-references are minimal and appropriate only at particular places in the text, be as specific as possible in telling the user why to see more information and where to find it, in that order. Exception: Avoid specific page or chapter number references to books in other documentation sets because they may change.

Correct

For information about using the Outlining toolbar, see "Starting an Outline," page 221.
For information about using the Outlining toolbar, see "Starting an Outline" in Chapter 17, "Outlining and Organizing a Document."
For information about using the Outlining toolbar, see "Starting an Outline" earlier in this chapter.

As a general rule, go from the specific to the general in the same cross-reference. That is, refer first to the section or topic heading, then the chapter number and title, and then the book title, using as many of these as necessary. If you use page numbers, however, do not put the page number first.

SEE ALSO **Bibliographies**

Style for cross-references

Follow these capitalization and punctuation guidelines for cross-references:

- Follow the capitalization style used in the titles and headings of your document, but as a general guideline use title caps for the names of books, chapters, sections, Web sites, and appendixes, including Help books and chapters in the Help contents. Exception: If you list an Internet address in printed documentation, follow the capitalization as it appears in the URL, which usually is lowercase.
- Italicize book titles in both printed and online documentation.
- Use quotation marks around titles of sections and chapters. Do not include the word *chapter* or its number within the quotation marks: Chapter 7, "Formatting."
- Capitalize the word *chapter, section, appendix, table,* or *figure* when it refers to a specific number, as in "see Chapter 7." Use lowercase when no number is mentioned, as in "earlier in this chapter."
- Capitalize *index, glossary,* and *appendix* in references such as "see the Index." Do not capitalize the word *page.*

SEE ALSO **Capitalization, Indexing**

Cross-references to art

Do not make cross-references to untitled or unnumbered art or tables unless the art or table immediately precedes or follows the reference on the page.

SEE ALSO **Art, Captions, and Callouts**

current drive

Not *current disk drive.*

current folder

Refers to the folder you are currently looking at (for example, in My Computer) or saving something to (for example, the folder that appears in the Save In box in the Save As dialog box).

current window

Do not use; use *active* or *open window* instead.

cursor

Do not use except in specific situations in programming documentation and references to MS-DOS-based programs. See your project style sheet for these correct uses. In all other documentation, use *pointer.* Do not use *mouse cursor* because pointers occur with other input devices as well.

SEE ALSO **insertion point**

cut

Do not use *cut* as a verb, even when referring to the Cut command. Do not use *cut* as a noun to refer to the Delete command (use *deletion*) or as an imperative in procedures (use *delete*). Do not use *cut-and-replace* or *cut-and-paste* as a verb or a noun.

Correct

Use the Cut command to delete the selected text.
Select the text you want to delete and then click Cut.

Incorrect

Cut the selected text.
Cut-and-paste the selected text.
Do a cut-and-paste on the second paragraph.

Do not use the verb *cut* to describe temporarily moving text to the Clipboard.

Correct

Use the Cut command to move the selected text to the Clipboard.

Incorrect

Cut the selected text to the Clipboard.

cut-and-paste

Okay to use as an adjective. Do not use as a verb or noun.

Correct

Perform a cut-and-paste operation.

Incorrect

Do a cut-and-paste.

Dangling Modifiers

A "dangling modifier," usually a verbal phrase, "dangles" because it does not logically or clearly modify what the writer intends. Dangling modifiers most often appear at the beginning of a sentence and often occur when the sentence is cast in the passive voice.

Dangling modifier

When playing audio that is written directly, it is difficult to avoid gaps.

Correct

When playing audio that is written directly, you [or "the user"] may encounter gaps.
When you play audio that is written directly, gaps can occur.

Here are some suggestions for fixing particular kinds of dangling modifiers.

Dangling gerunds

To correct a dangling gerund phrase, add or clarify the actor (a noun or pronoun) that the phrase logically modifies.

Dangling gerund phrase

By using object-oriented graphics, the structural integrity of the individual element of the graphic is maintained and can be edited.

Correct

By using object-oriented graphics, you can edit each element of the graphic because the structural integrity of the individual elements is maintained.

Dangling participles

To correct a dangling participial phrase, add a subject near the phrase or change the phrase to a clause with a subject.

Dangling participial phrase

Even after adding more data, the spreadsheet calculated as quickly as before.

Correct

Even after adding more data, the accountant thought the spreadsheet calculated as quickly as before.
Even though the accountant added more data, the spreadsheet calculated as quickly as before.

Dangling infinitives

To correct a dangling infinitive phrase, add a noun or pronoun to serve as the subject of the phrase.

Dangling infinitive phrase

To add original graphics to your document, a scanner is needed.

Correct

To add original graphics to your document, you need a scanner.

SEE ALSO *Harbrace College Handbook* and *Handbook of Technical Writing*

data

Use as either singular or plural in meaning but always with a singular verb. That is, always use "the data is" (or another appropriate verb) whether you mean a collection of facts (plural) or information (singular). If you want to emphasize that something is plural, rewrite to use a term such as *facts* or *numbers*. Do not use *datum* or *data are.*

Correct

The data shows that 95% of the users prefer a graphical interface.
The data gathered so far is incomplete.
These facts contradict the earlier data.

data record

Use just *record* instead.

database (adj, n)

One word.

datagram

One word. Refers to one packet, or unit, of information sent through a packet-switching network.

data modem

Two words. A modem that can both send and receive serial data. A data/fax modem can also send and receive faxes.

dates

Use this format to indicate a date in documentation: *month day, year,* as in January 31, 1999. Do not use *day month year* or an all-number method. Do not use ordinal numbers to indicate a date.

Correct

February 23, 1998
June 11, 1999

A B C D E F G H I J K L M N O P Q R S T U V W X Y Z

Incorrect

23 February 1998
6/11/98
April 21st

Avoid abbreviations of months unless necessary to save space.

datum

Do not use. See **data**.

deaf or hard-of-hearing

Use this phrase in accessibility information or if you refer to people who are deaf in your products. If the entire phrase is too long, use *deaf* alone. Do not use *hearing-impaired.*

SEE ALSO **Accessible Documentation**

debug (v)

Okay in technical documentation, but avoid in end-user documentation. Use *troubleshoot* or another accurate phrase instead.

decrement (v)

Means to decrease by one. Use only in technical documentation, but prefer this term to *decrease* when necessary for accuracy.

default (adj, n)

Avoid as a verb. It's jargon.

Correct (adjective)

If you don't select a template, Word uses Normal.dot, the default template.
This value is the context ID for the custom Help topic of the command. If it is omitted, the default Help context ID assigned to the macro is used.

Correct (noun)

If you don't select a template, Word uses Normal.dot by default.
This value is the number of sheets to add. If omitted, the default is one sheet.

Incorrect (verb)

If you don't select a template, Word defaults to Normal.dot.
This value is the number of sheets to add. If omitted, the program defaults to one sheet.

defragment (v)

Okay to use to refer to the action of the Disk Defragmenter program. Do not use *defrag.*

Correct

To defragment your files and speed up performance, use Disk Defragmenter frequently.

deinstall

Do not use except when necessary to refer to the interface; use *remove* instead.

SEE ALSO **uninstall**

delete (v)

Use *delete* both to refer to the Delete command and as an imperative in procedures. Do not use *cut* or *erase* as synonyms for *delete*.

Do not use *remove* to mean *delete*. Remove is correct, however, to refer to removing (not permanently deleting) items such as toolbar buttons or column headings in programs such as Outlook to customize an interface.

Correct

Delete the second paragraph.
Delete MyFile.txt from the Windows folder.
Remove the Size field from the Inbox.

SEE ALSO **cut, erase, remove**

depress

Do not use; instead, use *press* for the action of pushing down a key. Write around or otherwise avoid using *depressed* as a description for an indented toolbar button.

SEE ALSO **Key Names**

descriptor

Note spelling.

The descriptor refers to the type of software product that should follow a trademarked product name. For example, the descriptor for Developer Studio® is "visual development system."

deselect

Do not use; instead, use *cancel the selection,* or *clear,* in the case of check boxes.

desire

Do not use; use *want* instead.

desktop

Refers to the area within the program window. Use *client area* only if necessary in technical documentation. Valid for both Intel-based and Macintosh computers.

destination

General term for the name of the element you go to from a link, as well as other ultimate objects or ends. Acceptable to use in all documentation, but use "Web site," "page," or another term if it is more precise.

destination disk, destination drive, destination file

Not *target disk, target drive,* or *target file.* However, avoid if possible by using a word more specific to the context, as in "drag the folder to the icon for drive A or B."

device driver

Do not use in end-user documentation to refer to drivers in programs such as Microsoft Excel or Word that run on Windows 95 or Windows NT. Use the name of the specific kind of driver, such as *printer driver*, instead.

It's acceptable to use the term *device driver* when referring to the installation of peripheral devices such as disk drives, sound controllers. and so on.

dezoom

Do not use; use *zoom out* instead.

Dfs

Do not use all caps to refer to "distributed file system." Precede with *Microsoft* at first mention (Microsoft Dfs).

dialog

Do not use as an abbreviation for "dialog box." It's jargon. Do not use the spelling *dialogue.*

dialog box

In end-user and technical documentation, always use *dialog box,* not just *dialog* and not *pop-up window*—for example, *dialog box option* and *dialog box title.* In programming documentation, it's okay to use just *dialog* as an adjective with other entities such as *dialog class, dialog editor,* and *dialog object.*

Dialog Boxes and Property Sheets

Dialog boxes contain command buttons and various kinds of options through which users can carry out a particular command or task. For example, in the Save As dialog box, the user must indicate in which folder and under what name the document should be saved. (Toolboxes are simply dialog boxes with graphical options that are treated in the same way as other options.)

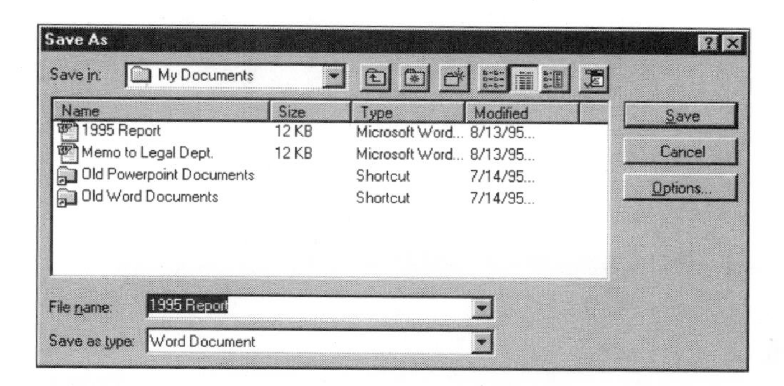

Property sheets display information ("properties") about an object in the interface. For example, the Taskbar property sheet shows information such as the size of the Start menu icons and whether the clock is shown on the taskbar. Property sheets have command buttons and, when properties can be edited, they can contain options, as dialog boxes do. Both dialog boxes and property sheets can have tabbed pages that group similar sets of options or properties.

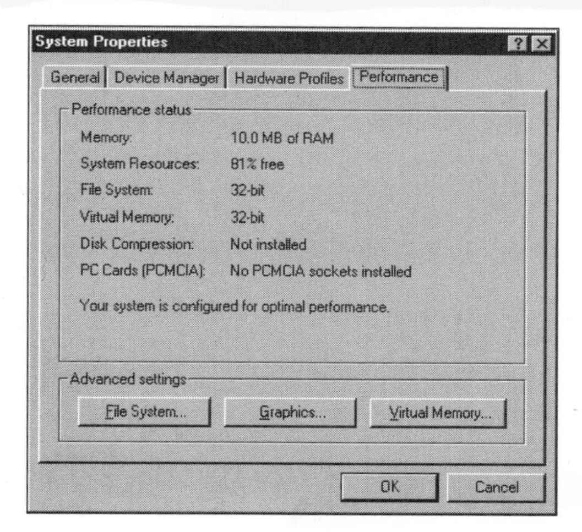

In most documentation, treat elements in dialog boxes and property sheets the same way. Avoid differentiating between property sheets and dialog boxes in end-user documentation. In general, avoid using the term *dialog box* or *property sheet* if you can and refer to property sheets as dialog boxes if you can't avoid a descriptor. Check your project style sheet, however. For programmer and administrator documentation, differentiate as necessary.

In programming and other technical documentation, buttons and other dialog box elements are called *controls,* especially in discussions about creating them. Do not use that term in end-user documentation.

A
B
C
D
E
F
G
H
I
J
K
L
M
N
O
P
Q
R
S
T
U
V
W
X
Y
Z

NOTE In some hardware products, buttons, switches, and so on are called "controls" because they give the user control over various actions. For example, users use joystick controls to move around the screen, especially in action games.

Dialog box syntax

These terms are most commonly used for user actions in dialog boxes:

- *Click:* Use for commands, command buttons, option buttons, and options in a list, gallery, or palette.
- *Select* and *clear:* Use for check boxes.
- *Type or select:* Use to refer to an item (as in a list box) that the user can either type or select in the accompanying text box. You can use *enter* instead if there's no possibility of confusion.
- *Choose* and *select:* Use these terms only when documenting generic procedures, not mouse procedures. Use *choose* for commands and *select* for options.

Except for the identifiers *box, list, check box,* and *tab,* the generic name of an item within a dialog box (*button, option,* and so on) should not follow the item's label, especially within procedures. *Check box* in particular helps localizers differentiate this item from other option boxes.

Use bold for dialog box titles, labels, and options.

The following example shows typical procedure wording for dialog box elements.

Correct

To view bookmarks

1. On the **Tools** menu, click **Options**, and then click the **View** tab.
2. Select the **Bookmarks** check box.

SEE ALSO **Procedures**

Dialog box elements

In most documentation, especially for end users, do not differentiate between elements such as drop-down combo boxes, list boxes, and so on. Do use the term *check box,* however. Use the correct label (for example, Save as type) with the term *box* or *list* if necessary to locate where the user should be and then direct the user to click, select, or take other action.

The following table describes the various elements that can appear in dialog boxes. In general, do not use the name of the particular element in documentation except in reference to designing interfaces. Use lowercase for the name of the element ("the Spaces check box"). In general, use sentence caps for the specific descriptor.

Element name	***Definition***	***Usage***	***Example***
Check box*	Square box that is selected or cleared to turn on or off an option. More than one check box can be selected.	Select the **Spaces** check box. Click to clear the **Bookmarks** check box. Select the appropriate check boxes. **NOTE** Always include *check box* with the label name.	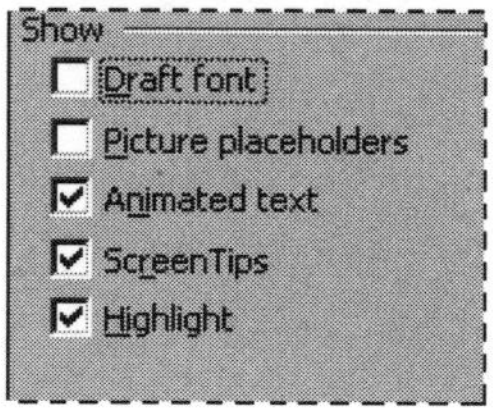
Combo box	Text box with a list box attached. The list is always visible. Because users can either type or select their choice, you can use enter to describe the action. Follow your project style sheet.	In the **Font** box, type or select the font you want to use. – or – In the **File Name** box, enter a file name.	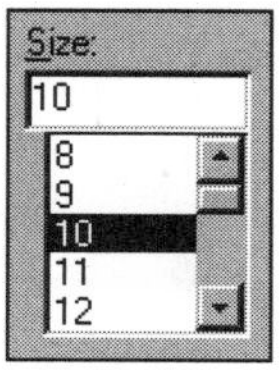
Command button	Rectangular button that initiates an action. A command button label ending with ellipses indicates that another dialog box will appear: More information is needed before the action can be completed.	Click **Options**.	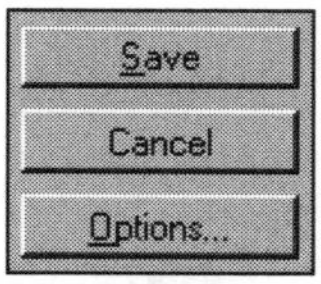
Drop-down arrow	Arrow associated with a drop-down combo or list box or some toolbar buttons, indicating a list the user can view by clicking the arrow.	Click the **Size** arrow to see more options.	
Drop-down combo box	Closed version of a combo box with an arrow next to it. Clicking the arrow opens the list.	In the **Size** box, type or select a point size.	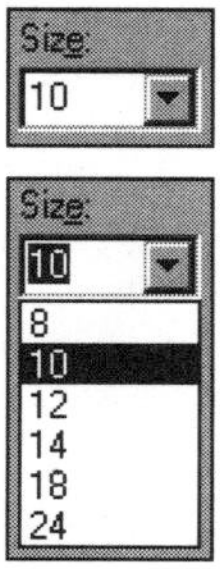

(continued)

Element name	*Definition*	*Usage*	*Example*
Drop-down list box	Closed version of a list box with an arrow next to it. Clicking the arrow opens the list. Depending on the type of list, use either *list* or *box*, whichever is clearer.	In the **Item** list, click **Desktop**.	Item: Desktop Item: Desktop Desktop Icon Icon Spacing (Horizontal) Icon Spacing (Vertical) Inactive Title Bar Inactive Window Border Menu
Group box	Frame or box that encloses a set of related options. It is a visual device only. If necessary for clarity, you can use either *under* followed by the label or *in the [name of group] area.*	Click **Small Caps**. – or – Under **Effects**, click **Small Caps**. – or – In the **Effects** area, click **Small Caps**.	Effects Strikethrough Superscript Subscript Hidden Small Caps All Caps
Label (do not use *caption*)	Text attached to any option, box, command, and so on. Refer to any option, box, and so on by its label.	In the **Font** list, click **Arial**.	Font:
List box	Any type of box containing a list of items the user can select. The user cannot type a selection in a list box. Depending on the type of list, use either *list* or *box*, whichever is clearer.	In the **Wallpaper** list, click the background wallpaper of your choice.	(None) Arcade Argyle Black Thatch Blue Rivets
Option button (avoid *radio button*)	Round button used to select one of a group of mutually exclusive options.	Click **Portrait**.	Orientation A Portrait Landscape
Slider* (also called *trackbar control* in some technical documentation)	Indicator on a gauge that displays and sets a value from a continuous range, such as speed, brightness, or volume.	Move the slider to the right to increase the volume.	Volume: Low High

Element name	*Definition*	*Usage*	*Example*
Spin box (do not use *spinner* or other labels)	Text box with up and down arrows that the user clicks to move through a set of fixed values. The user can also type a valid value in the box.	In the **Date** box, type or select the part of the date you want to change.	Start: 8 : 00AM End: 8 : 30AM
Tab* (also called *tabbed page* in technical documentation)	Labeled group of options used for many similar kinds of settings.	On the **Tools** menu, click **Options**, and then click the **View** tab. **NOTE** Always include *tab* with the label name.	Options Save Revisions View Normal Show Draft Font Wrap to Window Picture Placeholders
Text box	Rectangular box in which the user can type text. If the box already contains text, the user can select that default text or delete it and type new text.	In the **Size** box, select **10** or type a new font size. In the **Size** box, enter a font size. **NOTE** You can use *enter* if there's no chance of confusion.	10
Title (do not use *caption*)	Title of the dialog box. It usually, but not always, matches the title of the command name. Refer to the dialog box by its title when necessary, especially if the user needs to go to a new tab.	In the **Options** dialog box, click the **View** tab.	Options
Unfold button	Command button with two "greater than" signs (>>) that enlarges a secondary window to reveal more options or information.	Click **Profiles** for more information.	Profiles>>

* *Check box*, *tab*, and *slider* are the only terms in this table that should typically be used in end-user documentation.

SEE ALSO **Document Conventions**

A B C D E F G H I J K L M N O P Q R S T U V W X Y Z

dial-up (adj)

Use as an adjective only, not as a verb or noun. Always hyphenate.

As an adjective, it defines a line, a modem, or a networking connection. It refers to a service. Do not use as a noun ("a dial up"); it's ambiguous.

Use *dial* as the verb to refer to placing a call or using a dial-up device.

different

The adjective *different* is usually followed by *from.* Use *from* when the next element of the sentence is a noun or pronoun. There are constructions where "different to" or "different than" is correct or more idiomatic, but these are infrequent.

Correct

The result of the first calculation is different from the result of the second.
Her approach to programming is different from the others'.

Incorrect

The result of the first calculation is different than the result of the second.
Her approach to programming is different than the others'.

Do not use *different* to mean "many" or "various."

dimmed

Use *unavailable* to refer to commands and options that are in an unusable state, but use *dimmed* instead of *grayed* to describe the appearance of an unavailable command or option. (Use *shaded* to describe the appearance of check boxes that represent a mixture of settings.) Also, use *appears dimmed,* not *is dimmed.*

Correct

If the option appears dimmed, it is unavailable.

Incorrect

If the option is grayed, it is unavailable.

SEE ALSO **gray, grayed, shaded**

direction keys

Do not use; use *arrow keys* instead.

SEE ALSO **Key Names**

directory

In general, limit use of the word *directory* to references to the structure of the file system. Use *folder* to refer to the visual representation or object in the interface. You can include *directory* as a synonym for *folder* in indexes and search topics.

SEE ALSO **folder**

directory icon

Do not use; this term is no longer applicable. Use *folder icon* generically, if necessary.

disable

Acceptable in technical documentation in the sense of a programmer setting the guidelines for making a command unavailable. Avoid in other documentation. For example, in the sense of disabling a command, use *make unavailable* or something similar.

SEE ALSO **dimmed, Menus and Commands**

disabled

Do not refer to people with disabilities as *disabled*. See **Accessible Documentation**.

disc

Use only when referring to a compact disc (CD-ROM).

SEE ALSO **CD-ROM, disk**

discreet vs. discrete

Be sure to use these terms accurately. *Discreet* means "showing good judgment"; it's seldom needed in documentation. *Discrete* means "separate" or "distinct" and is more likely to appear in documentation.

disjoint selection

Do not use. Use *multiple selection* to refer to a selection of more than one item, such as options, or *nonadjacent selection* (not *noncontiguous selection*) to make it clear that the items are separated.

disk

In general, use *disk* to refer to both hard disks and floppy disks.

Unless necessary, use just *disk,* not *hard disk, floppy disk,* or *3.5-inch disk.* Do not use fractions or symbols when specifying a disk; use decimals and spell out *inch.*

Correct

3.5-inch disk

A B C D E F G H I J K L M N O P Q R S T U V W X Y Z

Do not use *diskette, fixed disk, hard drive,* or *internal drive.* Do not use *hard disk system* or *floppy disk system.* Refer to the computer specifications instead.

In general, do not use *disk* in possessive constructions, such as *disk's contents* or *disk's name;* instead, use *disk contents* or *disk name.*

When naming specific disks, use the disk names as they appear on the labels.

Correct

The utilities disk
Disk 1

NOTE Do not use *disk* to refer to a compact disc.

disk drive

Do not use *disk* alone to refer to a 5.25-inch or 3.5-inch disk drive.

SEE ALSO **drive**

disk icon

Use only with Macintosh products.

disk resource

Use to refer to a disk or part of a disk shared on a server.

disk space

Use *disk space,* not *storage* or *memory,* to refer to memory capability. *Storage device* is acceptable as a generic term to refer to things such as disk and tape drives.

diskette

Do not use; use *disk* instead.

SEE ALSO **disk**

display (n)

Use *display* as a noun to refer generically to the visual output device and its technology, such as a CRT-based display, a flat-panel display, and so on. Use *screen* to refer to the graphic portion of a monitor.

SEE ALSO **appears, displays**

display adapter, display driver

Avoid; use *video adapter* instead.

document

You can use *document* generically to refer to any kind of item within a folder that can be edited, but it's clearer to restrict its use to Word, WordPad, and text documents. Use the specific term for "documents" in other programs—for example, an Excel *worksheet,* a PowerPoint *presentation,* an Access *database.*

Use *file* for more general uses, such as *file management* or *file structure.*

Correct

These demos will help you learn how to manage files and folders, print your documents, and use a network.

Document Conventions

Consistent use of typographic conventions in documentation helps users locate and interpret information easily. The following guidelines present some specific typographic conventions.

NOTE Some elements may not appear here. Consult your project style sheet.

Item	*Convention*	*Example*
Accessory programs	Title caps	Heap Walker Nmake Notepad
Acronyms	All uppercase	CUA FIFO
A.M., P.M.	All uppercase	A.M., P.M.
Arguments (predefined)	Bold italic	**EditBookmark *.Name*** = *selectedTitle$,**.Add***
Attributes	Bold; capitalization varies	**IfOutputPrecision**
Book titles	Title caps, italic	See the *Visual Basic Custom Control Reference.*
Chapter titles	Title caps, in quotation marks	See Chapter 9, "Extending Forms."
Classes (predefined)	Bold; case varies	**ios** **filebuf**
Classes (user-defined)	Roman in text; case varies. Monospace in code samples.	The class money consists .... `class money`

Code samples, including keywords and variables within text and as separate paragraphs, and user-defined program elements within text	Monospace	`#include <iostream.h>`
		`void main ()`
		the pointer `psz`

(continued)

A B C D E F G H I J K L M N O P Q R S T U V W X Y Z

Item	*Convention*	*Example*
Command-line commands and options (switches)	All lowercase, bold	**copy** command **/a** option
Commands on menus and buttons	Bold; capitalization follows interface (usually title caps)	**Date and Time** **Apply** **New Query** button
Constants	Variable treatment; usually bold; case varies; always consult your project style sheet for treatment of constants	**INT_MAX** **dbDenyWrite** **CS**
Control classes	All uppercase	EDIT control class
Data formats	All uppercase	CF_DIB
Data structures and their members (predefined)	Bold; case varies	**BITMAP** **bmbits** **CREATESTRUCT** **hInstance**
Data types	Bold; case varies	**DWORD** **float** **HANDLE**
Device names	All uppercase	LPT1 COM1
Dialog box options	Bold; capitalization follows interface	Click **Close all programs and log on as a different user?** **Find Entire Cells Only** check box
Dialog box titles	Bold; title caps	**Protect Document** dialog box **Import/Export Setup** dialog box
Directives	Bold	**#if** **extern "C"**
Directories	Initial caps (internal caps are acceptable for readability)	\\Irstaxforms\Public \\IRSTax Forms\Public
Environment variables	All uppercase	INCLUDE SESSION_SIGNON
Error message names	Title caps	General Protection Fault
Extensions	All lowercase	.mdb .doc
Fields (Members)	Bold; capitalization varies	**lfHeight** **biPlanes**

Item	*Convention*	*Example*
File names	Title caps (internal caps in short file names are acceptable for readability)	My Taxes for 1995 Msacc20.ini MSAcc20.ini
Folders and directories	Title caps	My Documents Vacation and Sick Pay \\Accounting\Payroll\Vacpay
Functions (predefined)	Variable treatment; usually bold; case varies; always consult your project style sheet for treatment of functions	**CompactDatabase** **CWnd::CreateEx** **FadePic**
Handles	All uppercase	HWND
Hooks	All uppercase	WH_CBT
Icon names	Bold; title caps	**Recycle Bin** In Control Panel, click **Add New Hardware**.
Indexes	All uppercase	RASTERCAPS
Key names, key combinations, and key sequences	All uppercase	CTRL, TAB CTRL+ALT+DEL SHIFT, F7 ALT, F, O
Keywords (language and operating system)	Bold; case varies	**main** **True** **AddNew**
Logical operators	All uppercase, bold	**AND** **XOR**
Macros	Usually all uppercase; bold (if predefined); may be monospace if user-defined	**`LOWORD`** `MASKROP`
Members	Bold; lowercase, with internal caps	**ulNumCharsAllowed**
Members (Fields)	Bold; capitalization varies	**lfHeight** **biPlanes**
Member functions	Bold; capitalization varies	**Serialize** **CRuntimeClass::Store**
Menu names	Bold; title caps	**Insert** menu
New terms or emphasis (but note that italic does not always show up well online, so you can use quotation marks for new terms)	Italic	Microsoft Exchange consists of both *server* and *client* components. You *must* close the window before you exit.

(continued)

Item	Convention	Example
Operators	Bold	**+, -** **sizeof**
Options (command-line)	Bold; case-sensitive	**copy** command **/a** option **/Aw**
Parameters	Italic; capitalization varies from all lowercase to initial cap to lowercase with intermediate caps.	*hdc* *grfFlag* *ClientBinding*
Placeholders (in syntax and in user input)	Italic	*[form] Graph.***Picture** Type *password*
Programs and applications, including utility and accessory programs	Usually title caps (check the Microsoft trademark list for other styles of capitalization)	Microsoft Word Notepad Dial-Up Networking Lotus 1-2-3 for Windows Microsoft At Work Heap Walker Nmake
Properties	Variable treatment; usually bold; always consult your project style sheet for treatment of properties	**M_bClipped** **AbsolutePosition** **Message ID**
Registers	All uppercase	DS
Registry settings	Subtrees (first-level items): all uppercase, separated by underscores; usually bold. Registry keys (second-level items) either all uppercase or mixed case (no underscore); Registry subkeys (below the second level) usually mixed case and bold, but check your project style. See **registry**.	**HKEY_CLASSES_ROOT** **HKEY_LOCAL_MACHINE** **SOFTWARE** **ApplicationIdentifier** **Microsoft**
Statements	All uppercase; bold	**IMPORTS** **LIBRARY**
Strings	Sentence caps; enclosed in quotation marks	"Now is the time"
Structures	All uppercase; usually bold	**ACCESSTIMEOUT**
Switches	Usually lowercase; bold	**build: commands**
System Commands	All uppercase	SC_HOTLIST
Toolbar button names	Usually title caps (follow the interface); bold	**Format Painter** **Insert Microsoft Excel Worksheet**

Item	*Convention*	*Example*
URLs	All lowercase; break long URLs before a forward slash, if necessary to break; do not hyphenate. The protocol name can be omitted when telling someone to connect. See **URL**.	http://www.microsoft.com/ seattle.sidewalk.com /music/ //www.microsoft.com/
User input	Usually lowercase (note also bold or italic, depending on element), unless case-sensitive or to match standard capitalization conventions	Type -**p***password*
Utilities	Usually title caps	Makefile RC program
Values	All uppercase	DIB_PAL_COLORS
Variables	Variable treatment; always consult your project style sheet for treatment of variables	bEmpty **m_nParams** *file_name*
Views	Usually lowercase, but varies by product. Capitalize those based on proper names (Gantt view).	outline view, chart view, but Design view (Access). One guideline for groups to use for deciding is to follow the interface for views listed as menu commands.
Windows, named	Title caps	Help window
Windows, unnamed	All lowercase	document window

SEE ALSO **Capitalization, Code Formatting Conventions, Command Syntax, File Names and Extensions, font and font style, Key Names, Menus and Commands**

domain

Because *domain* has different meanings in database design, Windows NT, and Internet addresses, define the use or make sure the context is clear. Always consult your project style sheet.

In database design, a domain is the set of valid values for a particular attribute. In Windows NT, a domain is a collection of computers sharing a common database and security policy. And on the Internet, the domain is the last part of the address, following the dot. It identifies the type of entity owning the address (such as .com for commercial entities).

Current Internet domains are .com, .edu, .gov, .mil, .org, and .net. More domains will be added soon, however.

For more detailed definitions of domain, see the *Microsoft Press Computer Dictionary*.

done

Do not use *when you are done;* it's colloquial. Use *when you have finished* instead.

A B C D E F G H I J K L M N O P Q R S T U V W X Y Z

DOS

Acronym for *disk operating system;* avoid. Whenever discussing the product or the disk operating system in general, use *MS-DOS*. See **MS-DOS**.

dotted rectangle

Avoid; use this term only if you are graphically describing the element that a user drags to select a region on the screen. Use ***bounding outline*** (not *marquee*) instead.

double buffering (n)

No hyphen. Refers to the use of two temporary storage areas. Do not use as a verb. Use a phrase such as "uses double buffering" instead.

double-click, double-clicking

Always hyphenate. Use instead of *select* and *choose* when referring to a mouse action. Do not use *double-click on.*

double word

Two words. Refers to a unit of data consisting of two contiguous words (bytes). DWORD is used in code.

download

Acceptable to refer to the process of transferring a copy of a program or file from the Internet or a remote computer or server or a personal computer.

drag

Avoid *click and drag* or *press and drag.* See **mouse**.

drag-and-drop (adj)

Do not use as a verb and avoid using as a noun. The action of dragging includes dropping the element in place.

Okay to use as an adjective to describe the feature of moving objects between windows and programs and to describe behavior a programmer wants to put in a program. In these cases, use a phrase such as "drag-and-drop editing," "a drag-and-drop feature," and so on.

Correct

Moving files is an easy drag-and-drop operation.
You can drag the folder to drive A.
You can move the folder to drive A using a drag-and-drop operation.

Incorrect

You can drag and drop the folder in drive A.
You can use drag-and-drop to move the folder to drive A.
Drag the information from Microsoft Excel and drop it in a Word document.

drilldown (n), drill down (v)

Do not use in documentation to refer to following a path to its files or to further analysis. It's slang.

drive

Distinguish among types of disks and disk drives only when necessary to avoid confusion.

Use these conventions when referring to drives:

- Use *drive, disk drive,* or, if necessary, *floppy disk drive* to describe the 3.5-inch disk drives in a computer. Do not use *disk* alone to refer to a disk drive.
- Use *hard disk*, not *hard drive* or *hard disk drive*, to refer to the hard disk itself.
- Use *CD-ROM drive.*
- Use *current drive,* not *current disk drive* or *active drive.*
- Use *drive A,* not *drive A:, drive A>,* or *A: drive.*
- Use *network drive* to refer to a logical network drive name, such as *network drive X.*

Because Macintosh drives don't have names, it's acceptable in introductory material about Macintosh computers to describe a drive (as in "if you have two disk drives, use the one on the right") to be consistent with Macintosh documentation.

Avoid telling users to "close the drive" unless you're writing a section introducing computers.

drive name

Not *drive specification, designator,* or *designation.*

drop-down (adj)

Use only if necessary to describe how an item such as a menu works or what it looks like. Acceptable in technical documentation if necessary to describe the type of item, as in *drop-down arrow, drop-down combo box,* or *drop-down list box.*

SEE ALSO **Dialog Boxes and Property Sheets, Menus and Commands**

dynamic-link library

Abbreviate as *DLL* after first mention. If you are sure your audience knows the term and that localization will not be affected, you can use *DLL* without spelling it out. Use lowercase (.dll) when referring to the file extension. Do not use *dynalink.*

earlier

Use instead of *above* in cross-references—for example, "earlier in this section."

Use instead of *lower* for product version numbers—for example, "Word version 3.0 or earlier."

SEE ALSO **Cross References, Version Numbers**

edit

Because the term can be confused with the Edit menu, avoid it in end-user documentation to refer to making changes in a document; use *change* or *modify* instead.

edutainment

Do not use in documentation. It's marketing jargon to refer to educational software (usually multimedia or Web-based) that purports to entertain while it educates. The term can also cause difficulties for localization.

e.g.

Means *exempli gratia.* Do not use; use *for example* instead.

8.5-by-11-inch paper

Not 8.5 x 11-inch, 8 1/2 by 11-inch, or other ways of referring to the paper size.

either/or

Do not use. Fill out the construction, as in "you can either close the document or quit the program."

Ellipses

In general, avoid using an ellipsis (...) except in syntax lines or to indicate omitted code in technical material.

An ellipsis is typically used in the interface to show truncation, as in a program name, or to indicate on menus and in dialog boxes that more information is required from the user to complete the command. Do not use an ellipsis in this context in documentation.

In printed material, you can use an ellipsis in multiple-part callouts, especially with a screen shot used in a procedure.

Correct

Click here ... ... and then move the pointer.

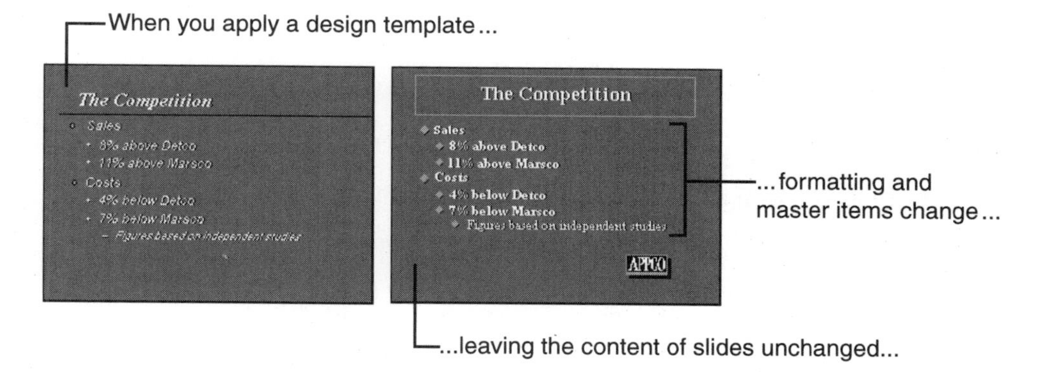

In callouts, close up ellipsis points, but insert one character space before and/or after them, as shown in the preceding example. If the sentence or phrase ends with additional punctuation such as a period or comma, insert the ellipsis one space after the punctuation mark.

Em Dash

The em dash (—), based on the width of an uppercase *M,* is used primarily to set off sentence elements.

> **NOTE** Insert a hairline space before and after an em dash if your style sheet and publishing process supports it. Do not use word spacing on either side of an em dash.

Use an em dash to:

- Set off within a sentence a parenthetical phrase that deserves more emphasis than parentheses imply. Use two em dashes, one on each side of the phrase.

  Correct

  The information in your spreadsheet—numbers, formulas, text—is stored in cells.

- Set off a phrase at the end of a sentence for emphasis. Use one em dash.

  Correct

  Set key names in all caps—for example, CTRL or ALT.

Do not use an em dash in place of a bullet or other typographic symbol to set off items in a list.

When a complete sentence follows an em dash, do not capitalize the first word unless it's a proper noun.

The HTML code for an em dash is —.

A B C D E F G H I J K L M N O P Q R S T U V W X Y Z

E-Words

In general, avoid forming new words with "e-" (for *electronic*) unless you know your audience will understand. Some words that may be appropriate in certain circumstances are "e-commerce" and "e-money." **E-mail** and **e-form** are acceptable. Use lowercase and always hyphenate for clarity.

e-form

Okay to use. Spell out as *electronic form* on first use if necessary for your audience.

e-mail (adj, n)

Okay to use to refer to an electronic mail program, as in "check your e-mail for messages," but use *e-mail messages,* or just *messages* or *notes,* to refer to pieces of e-mail. Do not use *e-mails.*

Avoid as a verb, as in "e-mail the file." Instead use *send* or *send in e-mail.*

Maintain the hyphenation to show the meaning of "electronic mail" and to be consistent with terms such as "e-commerce." Use *E-mail* at the beginning of a sentence and in headings.

e-zine, webzine

Avoid except to connote an underground-type of electronic magazine. It's okay to use *webzine* to refer to mainstream magazines such as *Slate* or *PC Week* that are on the Web, but it's better to call them *electronic magazines* or, if the electronic context is clear, just *magazines.*

embed

Not *imbed,* which is a variant spelling.

En Dash

The en dash (–) is half the length of an em dash (that is, the width of an *N*) and slightly longer than a hyphen. It is used primarily as a connecting element, especially with numbers.

Use an en dash:

- To indicate a range of numbers such as inclusive values, dates, or pages.

  Correct
  © 1993–1994
  pages 95–110

- To indicate a minus sign.
- To indicate negative numbers: –79.
- Instead of a hyphen in a compound adjective in which at least one of the elements is an open compound (such as "Windows 95"). That is, don't mix hyphens and en-dashes in the same compound adjective.

Correct
dialog box–type options
Windows NT–based programs
non-Windows-based programs [hyphenated]
MS-DOS-compatible products [hyphenated]

Incorrect
MS-DOS–compatible products
non–Windows-based programs

Do not use spaces on either side of an en dash.

The HTML code for an en dash is –.

enable, enabled

Acceptable in technical documentation in the sense of a programmer setting the guidelines for making a command available.

Avoid in other documentation. For example, in the sense of enabling a command, use *make available* or something similar. In the sense of telling the user something is possible, use *you can* instead. However, you can use *enable* or *allow* in instances where you must use the third person, as in "Microsoft Exchange allows a user to log on as a guest" or "the function enables the program to accomplish something." But avoid anthropomorphism.

SEE ALSO **allow, can vs. may**

end

As a verb, use to refer to stopping communications and network connections. Use *quit* for programs.

Correct

To end your server connection, on the Tools menu, click Disconnect Network Drive.

end user (n), end-user (adj)

Avoid; use *user, customer,* or *you* instead. Okay to use *end user,* however, to differentiate between types of users when, for example, there are both an administrator and end users or when referring to a programmer's audience in programming documentation.

endline (adj)

One word, as in *endline comment.*

endpoint

One word. Refers to the beginning or end of a line segment.

ensure

Means "to make certain" or "guarantee." Do not confuse with *insure,* which refers to insurance, or *assure,* which implies giving positive information.

enter (v)

Do not use as a synonym for *type* except to indicate that a user can either type or click a selection from, say, a list in a combo box.

Correct

Type your password, and then press ENTER.
In the File name box, enter the name of the file.

Incorrect

Enter your password, and then click OK.
At the prompt, enter the path and file name.

enterprise

Okay to use in client/server documentation to mean "large company" or "corporation." Use as an adjective if possible, as in "enterprise computing" or "enterprise networking" rather than as a noun to mean "corporation." Avoid in end-user documentation.

entry

Do not use as a synonym for *topic* in reference documentation.

entry field

Do not use in end-user documentation to refer to a text-entry field in a dialog box; use *box* or *text box* instead. Okay to use in programming or database documentation or where you may need to refer to some sort of data entry.

environment variable

Do not use *environment setting* or *environment string*.

erase

Do not use as a synonym for *delete.* It's okay to use *erase* for specialized purposes when the program requires it, as in Paint.

Error Messages

For most documentation, use *messages* instead of *error messages.* Okay to use if necessary in technical documentation to distinguish from other types of messages. Do not use *error code*. Use *error value* or *return value* instead.

Writing error messages

When you write error messages, use the passive voice to describe the error and, if necessary, the third person (refer to "the computer" or "the program"). You can also blame the product. Addressing the user directly as "you" may imply that the user caused the problem. Try to make error messages friendly, direct, and helpful.

Correct

The name of the object file for %s conflicts with that of another program in the project.
Works did not find a match for this term.

Incorrect

You have named the object file for %s with the name of another program in the project.
The term you typed does not appear in this Works file.

If possible, construct an error message so that the string that appears to the user briefly offers a solution. If that's not possible, the message should include a button leading to Help on that message.

Useful message

502 Authentication error. Error authorizing your password with the News Server.
Try an anonymous connection by removing your user name and password from the Communications dialog box.

This message gives the compiler string, which is useful to programmers. The additional information gives users a possible way to solve the problem.

Also, the OK button on most error messages raises users' hackles. Although some user action is required to close the message box, use a more friendly or logical term on the command button, such as "Close," if possible.

SEE ALSO **Messages**

et al.

Abbreviation for *et alia.* Do not use, except in a text reference citation of three or more authors; use *and others* instead.

etc.

Abbreviation for *et cetera.* Do not use; use *and so forth* or *and so on* instead.

executable file

A file with an .exe or .com extension. In nontechnical documentation, use *program file* instead. Use *executable* and *.exe* as adjectives only, never as nouns. Use the article *an,* not *a,* with .exe, as in "an .exe file."

Correct

an executable program
the .exe file

A B C D E F G H I J K L M N O P Q R S T U V W X Y Z

Incorrect

an executable
the .exe

execute (v)

Avoid in end-user documentation; for commands, use *carry out;* for programs and macros, use *run*.

Acceptable in programmer documentation because it has become ingrained, especially in the passive voice, as in "Commands are executed in the order they are listed in the file." However, *run* can often be accurately substituted for *execute* even in programmer materials.

SEE ALSO **run vs. execute**

exit

Use as the command name only; use *quit* as a transitive verb, as in "quit the program."

Correct

To quit, click Exit Works.

expand, collapse

Pertains to a folder or outline. The user can *expand* or *collapse* these structures to see more or fewer subentries. A plus sign next to a folder indicates that it can be expanded to show more folders; a minus sign indicates that it can be collapsed.

Explorer

Always use *Windows Explorer* or *Microsoft Internet Explorer,* depending on the program; do not shorten to *Explorer* or *the Explorer.*

expose

Do not use. Use a term such as "make available" or "display" instead.

extend

In the sense of extending a selection, use instead of *grow.*

extension, file name extension

Not *file extension*—for example, the .bak extension. See **File Names and Extensions**.

facsimile (n)

Do not use to refer to the kind of document sent through a fax machine; use *fax* instead. Use *facsimile* only to refer to an exact reproduction of something else.

SEE ALSO **fax**

fail

In end-user documentation, use only to refer to disks. Use *stop responding* to refer to programs. Do not use *crash.*

In programmer documentation *fail* often must be used. For example, E-FAIL is a common status return value that may require a discussion of failed items.

Correct

Backing up your files safeguards them against loss if your hard disk fails.

SEE ALSO **crash**

FALSE

All caps as a return value in programmer documentation.

family

Use instead of *line* to refer to a series of related Microsoft products.

Far East

Do not use. Use *Asia* or *Asian* instead.

far-left, far-right

Avoid; use *leftmost* or *rightmost* instead. If possible, however, avoid directional cues for reasons of accessibility. See **Accessible Documentation**.

fast key

Do not use; use *shortcut key* instead. See **Key Names**.

favorite

Reference in Internet Explorer to a Web page or site the user may want to return to. Favorites can be added to the menu. Corresponds to "bookmark" in other browsers. Use lowercase when referring to a "favorite Web site" and uppercase when referring to the Favorites menu.

Correct

You can add a favorite Web site to the Favorites menu.
You can display your list of favorites at any time by clicking the Favorites menu.

fax (n, adj, v)

Abbreviation for *facsimile.* Okay to use as a noun ("your fax arrived"), an adjective ("fax machine," "fax transmission"), or a verb ("fax a copy of the order"). Do not use *FAX.*

field

Do not use to refer to a box or option in a dialog box. Okay to use to refer to Word field codes and for other technically accurate uses.

Figure

Use for numbered art, as in *Figure 5.2.* Always capitalize in such uses. See **Art, Captions, and Callouts**.

file

Okay to use generically to refer to documents and programs as well as to units of storage or file management. However, be more specific if possible in referring to a type of file—for example, the Word *document,* your *worksheet,* the WordPad *program,* and so on.

file attributes

Use lowercase for the MS-DOS file attributes *hidden, system, read-only, archive,* and so on.

file extension

Do not use; use *extension* or *file name extension* instead.

file name (adj, n)

Two words. See **File Names and Extensions**.

A B C D E F G H I J K L M N O P Q R S T U V W X Y Z

file name extension, extension

Not *file extension.* See **File Names and Extensions**.

File Names and Extensions

For new and revised documentation for Windows and the Macintosh, use title caps for file names and directory and drive names. Use all lowercase for extensions.

Correct

.com file
C:\Taxes\April98
My Tax File, 1998

You can use internal caps for readability in concatenated file names if they won't be confused with function names, as in "MyTaxFile."

If the documentation is in transition—that is, if not all of it is being revised, or the schedule doesn't allow for change—use the previous file name convention of all uppercase, as in "C:\TAXES\APRIL98.XLS."

Do not use the term *file extension;* use *file name extension* or *extension.* If possible, write around the term in end-user documentation: "Word adds .doc to the file name if you don't specify the file format."

In general, however, avoid the use of any extension in end-user documentation. Describe the type of program or file instead (*an application, a text document, a worksheet,* and so on).

Precede file name extensions with a period. Use the article (*a* or *an*) that applies to the sound of the first letter of the extension, as though the period (or "dot") is not pronounced, as in "a .com file" and "an .exe file."

When instructing users to type literal file names and paths, use all lowercase bold for literals. Use italic for placeholders.

Correct

In the Command-line box, type ...
c:\msmail\msmail.mmf *password*

> **NOTE** Do not use *Foo, Fu, Foo.bar,* and the like as a placeholder for a file in Microsoft publications. Use a substitute such as *Sample File* or *MySample* instead.

SEE ALSO **Capitalization, Document Conventions**

finalize

Do not use; use *finish* or *complete* instead.

find and replace (adj, v)

Microsoft applications use *find* and *replace* as standard names for search and substitution features. Do not use *search and replace.*

Use *find* and *replace* as separate verbs. Do not use *find and replace* as a noun. Also avoid *search your document;* use *search through your document* instead.

Correct

Find the word "gem" and replace it with "diamond."
Search through your document, and replace "cat" with "dog."

Incorrect

Do a find and replace.
Find and replace the word "gem" with the word "diamond."

Avoid the term *global* in reference to finding and replacing unless absolutely necessary. This technical jargon may not be clear to all users.

Correct

Click Replace All to find all occurrences of the word "gem" and replace them with "diamond."
Replace all instances of the word "gem" with "diamond."

Use *find characters* and *replacement characters* to specify what the user types into a find-and-replacement text box.

Finder

Use *the Finder* only with Macintosh products. Always precede with *the.*

finished

Use instead of *done,* as in *when you have finished,* not the colloquial *when you are done.*

firewall

Combination of hardware and software that provides a security system, usually to prevent unauthorized access from outside to an internal network or intranet.

Avoid in end-user documentation. Define the term if you do use it there.

First Person

Avoid the first person (*I* and *we*) except in some marketing or legally oriented sections of books or Help or in troubleshooting sections. There, it is acceptable to say "we recommend" or "we suggest" to encourage the user to take some action, such as sending in the registration card or keeping purchase records. The first person in these constructions is friendlier than the passive "it is recommended." Never use *I* except when writing as a character such as Clippie ("Hi, I'm Clippie, and I'm your guide to Microsoft Office").

Correct

We recommend keeping the product disks, the Certificate of Authenticity, and your purchase receipt.

fixed disk

Do not use; use *hard disk* instead.

floppy disk

Use *disk* unless you need to distinguish between *floppy disk* and *hard disk.* Do not use *floppy drive* or *floppy disk system.*

Do not use *floppy* alone as a noun to refer to a disk. It's slang.

flush (v, adj)

In end-user documentation, do not use *flush, flush left,* or *flush right* to describe text alignment; use *even, left-aligned,* or *right-aligned,* as appropriate, instead.

In programmer documentation, *flush* is acceptable in contexts such as referring to a function that "flushes the buffer."

flush to

Avoid; use *aligned on* instead.

folder

In general, for graphical programs from Windows 95 and later, use *folder* to refer to what used to be called a directory or subdirectory. Folders are represented on the interface by a folder icon, and the term is based on the analogy that a folder stores other folders and documents of various types. Reserve *directory* for specific references to the structure of the file system, particularly in nongraphical programs such as MS-DOS and in programming documents.

> **NOTE** Not all folders replace directories. For example, the Printers and Control Panel programs are also folders. Describe the nature of the folder, if necessary.

Correct

You can find the file on your hard disk in \\Windows\System\Color.
You can find the file on X:\\Windows\System\Color.
You can find the file in the Color folder.
The system files are in the System subdirectory in the Win95 directory.

SEE ALSO **directory**

folder icon

Not *directory icon.*

A B C D E F G H I J K L M N O P Q R S T U V W X Y Z

following

Use *following* to introduce art, a table, or, in some cases, a list.

Correct

The following table compares the different rates.
To install the program, do the following:

If *following* is the last word before what it introduces, follow it with a colon.

SEE ALSO **above, below**

font and font style

Use *font,* not *typeface,* for the name of a typeface design—for example, Times New Roman or Bookman. Use *font style,* not *type style,* to refer to the formatting, such as bold, italic, or small caps, and *font size,* not *type size,* for the point size, such as 12 point or 14 point.

When referring to bold formatting, use *bold,* not *bolded, boldface,* or *boldfaced.* When referring to italic formatting, use *italic,* not *italics* or *italicized.*

Correct

The Bold option makes selected characters bold or removes the bold formatting if the characters are already bold.

Incorrect

Select the Bold option button to bold the characters.

For information on when to use various font styles, see **Document Conventions**.

foobar, fubar

Do not use; the word is slang derived from an obscene phrase meaning "fouled up beyond all recognition." Use another placeholder file name instead—for example, *Sample* or *MyFile.doc.*

footer

In word-processing and publishing documentation, use instead of *bottom running head* or *running foot* when discussing page layout; however, *running foot* is acceptable if needed for clarification or as a keyword or index entry.

footnotes

Avoid footnotes in normal text; however, footnotes are acceptable in tables. For use of footnotes in tables, see **Tables**.

foreground program

Not *foreground process.*

Foreign Words and Phrases

Avoid non-English words and phrases, such as *laissez-faire* or *ad hoc,* even if you think they're generally known and understood. They may not be, or the language may not be understood by a translator. Find a straightforward substitute in English instead.

Do not use Latin abbreviations for common English phrases.

Use these terms	Instead of
and so on	etc.
carefree	laissez-faire
for example	e.g.
improvised	ad hoc
that is	i.e.

fork

One of two parts of a Macintosh file. Each file has a resource fork and a data fork. The resource fork contains reusable types of information such as fonts, windows, menus, and so on. The data fork contains user-supplied information. For more information, see the *Microsoft Press Computer Dictionary.*

format, formatted, formatting

Use *format* to refer to the overall layout or pattern of a document. Use *formatting* to refer to particulars of character formatting, paragraph formatting, and so on. Note spelling.

Fortran

Do not use FORTRAN (all uppercase) to refer to the programming language, and never spell out its origin of Formula Translation.

4GL

Abbreviation for "fourth-generation language." Spell out at first mention.

frame

Avoid in end-user documentation to refer to a nonbrowser section of a Web page. Also avoid in technical documentation unless you are specifically referring to frame technology, not just a section of the page. (Many Web sites use tables, not frames, to divide a page.)

"Frame" has a number of other computer meanings (see the *Microsoft Press Computer Dictionary*), so be sure to define it if the context is unclear.

friendly name

Do not use. Use *display name* instead to refer to a person's name as it appears in an address or e-mail list.

A B C D E F G H I J K L M N O P Q R S T U V W X Y Z

from

Use *from* to indicate a specific place or time as a starting point: "Paste the text from the Clipboard" or "From the time you set the clock, the alarm is active."

Use *from* to indicate a menu from which a user chooses a command if you are documenting both mouse and keyboard procedures: "From the File menu, choose Open." Use *on,* however, to indicate the starting place for clicking a command or option: "On the File menu, click Open."

from vs. than

The adjective *different* is usually followed by *from.* Use *from* when the next element of the sentence is a noun or pronoun.

Correct

The result of the first calculation is different from the result of the second.

Incorrect

The result of the first calculation is different than the result of the second.

front end (n), front-end (adj)

Avoid as a synonym for the desktop interface to a database or server; it's marketing jargon. Instead, use the name of the program or *interface, application,* or another specific and accurate term.

FrontPage

A Microsoft product for Web authoring and management. Note capitalization. Trademarked term.

function

A general term for a subroutine. In some languages, a function specifically returns a value, which not all subroutines do.

Do not use *function* with API. Because an API is a set of functions, "API function" is redundant.

For more information about functions, routines, subroutines, and procedures, see the *Microsoft Press Computer Dictionary.*

G, G byte, Gbyte

Do not use as an abbreviation for *gigabyte*. Use *GB* instead.

SEE ALSO **GB, gigabyte**

gallery

A gallery is a collection of pictures, charts, or other graphics that the user can select from. Refer to the items in a gallery in the same way you refer to options in a dialog box—that is, use the verb *click* or *select*.

Correct

Select an option from the gallery.
Click the picture you want to select.

Gallery of clip art

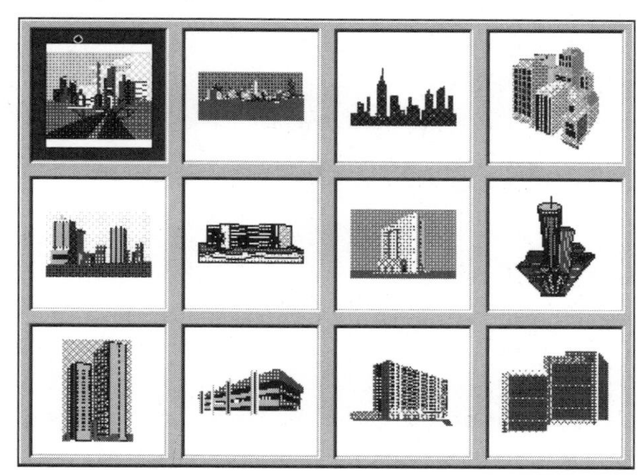

A B C D E F G H I J K L M N O P Q R S T U V W X Y Z

game pad

Two words and lowercase, even to refer to the Microsoft product—for example, Microsoft SideWinder game pad.

garbage collection

Even though it's jargon, this is a commonly used and acceptable term in programming documentation. It refers to the automatic deletion of objects that the environment perceives are no longer being used or to the automatic recovery of heap memory. A more formal but much less understandable synonym is "unpredictable finalization."

gateway

One word. Refers to software or a computer running software that enables two different networks to communicate.

GB

Abbreviation for *gigabyte.* Use the abbreviation only as a measurement with numerals; do not use in straight text without a numeral. Spell out *gigabyte* at first mention. See **gigabyte**.

Gbit

Do not use as an abbreviation for *gigabit;* always spell out.

general protection fault

Acceptable to abbreviate as *GP fault* after first mention.

Gerunds

In general, use gerunds (the *ing* form of a verb used as a noun) in book and chapter-level headings in Help tables of contents.

Correct

Managing Hardware and Software
Installing New Software

Incorrect

How to Install New Software

Note, however, that gerunds in text can sometimes be ambiguous, causing problems for localization. Be sure all necessary words are included to make the meaning clear.

Correct

You can change files by using the Template utility.
You can change files that use the Template utility.

Ambiguous

You can change files using the Template utility.

SEE ALSO **Headings and Subheadings, International Considerations, Procedures**

GHz

Abbreviation for *gigahertz.* Use the abbreviation only as a measurement with numerals; do not use in straight text without a numeral. Spell out *gigahertz* at first mention. See **gigahertz**.

gigabit

Always spell out. Do not use the abbreviation *Gbit.*

gigabyte

One gigabyte is equal to 1,073,741,824 bytes, or 1,024 megabytes.

- Abbreviate as *GB,* not *G, G byte*, or *Gbyte.* At first mention, spell out and use the abbreviation in parentheses.
- Leave a space between the numeral and *GB* except when the measurement is used as an adjective preceding a noun. In that case, use a hyphen.

  Correct
  1 gigabyte (GB) of data
  10-GB hard disk

- When used as a noun in measurements, add *of* to form a prepositional phrase.

  Correct
  You will need to free 1 GB of hard disk space.

SEE ALSO **Measurements**

gigahertz

A gigahertz is a unit of frequency equal to 1 billion cycles per second.

- Abbreviate as *GHz.* At first mention, spell out and use the abbreviation in parentheses.
- Leave a space between the numeral and *GHz* except when the measurement is used as an adjective preceding a noun. In that case, use a hyphen.

  Correct
  a frequency of 11.9300 gigahertz (GHz)
  11.9300-GHz communications

SEE ALSO **Measurements**

A B C D E F G H I J K L M N O P Q R S T U V W X Y Z

given (adj)

Do not use to mean *specified, particular,* or *fixed.*

Correct

Look in the specified folder.
Use the Find command to search for all occurrences of a specific word.
The meeting is always at a particular time.

Incorrect

Look in the given folder.
Use the Find command to search for all occurrences of a given word.
The meeting is always at a given time.

global

In technical documentation, acceptable to refer to memory that is accessible to more than one process, to a variable whose value can be accessed and modified by any statement in a program (called a "global variable"), and to similar elements that pertain to an entire program.

Avoid in end-user documentation, especially in describing find and replace procedures. Instead, describe the action being taken.

Correct

A cascading style sheet establishes global design formats.
Use the Find and Replace commands to find all occurrences of specific text and replace it with different text.

SEE ALSO **find and replace**

glyph

Avoid. This is jargon when used to refer generically to a graphic or pictorial image on a button, on an icon, or in a message box. Use *symbol* instead. Okay to use in a technical discussion of fonts and characters.

SEE ALSO **icon, symbol**

GP fault

Abbreviation for *general protection fault.* Spell out at first mention.

graphic, graphical, graphics

As a noun, use *graphic* to refer to a picture, display, chart, and so on.

As an adjective, use *graphics* to refer to software and *graphic* to mean "vivid" or "realistic" or to refer to the field of graphic arts. Use *graphical* to refer to an environment.

Noun

To import a graphic from another file, click Picture.

Adjective

Select the graphics file you want to open.
The image is graphic and accurate.
A tutorial offers the basics in graphic design.
The graphical user interface simulates a coliseum.

graphics adapter

Avoid; use *video adapter* instead.

SEE ALSO **video adapter**

gray, grayed

Do not use to refer to commands or options; use *unavailable* or *dimmed* instead. If you need to describe the appearance of check boxes with portions of a larger selection that are already selected, use *shaded,* not *grayed.*

Correct

In the Effects group box, names of selected options may appear shaded.
The Print command on the File menu is unavailable.

SEE ALSO **dimmed, shaded, unavailable**

grayscale (adj, n)

One word.

greater, better

Do not use either term to designate system requirements or versions of a program. Use *later* instead.

Correct

The program runs on Windows 3.1 or later.
You need a 486 or later processor.

Incorrect

The program runs on Windows 3.1 or greater.
You need a 486 or better processor.

SEE ALSO **higher**

gridline

One word.

A B C D E F G H I J K L M N O P Q R S T U V W X Y Z

group, newsgroup

Although these words can be synonyms, use *newsgroup* to refer specifically to an Internet newsgroup to differentiate it from other generic groups. *Newsgroup* is one word.

group box

Two words, lowercase. A group box is a standard control used to group a set of options in a dialog box. For example, page ranges are grouped in the Page range group box in the Print dialog box in Word.

The term is acceptable in programmer documentation. Avoid in end-user documentation. To save space, it's generally unnecessary to include the name of the group box in procedures unless options with the same label appear in more than one group box. In that case, use *under* with the box name, as in "Under Effects, click Hidden."

Group box

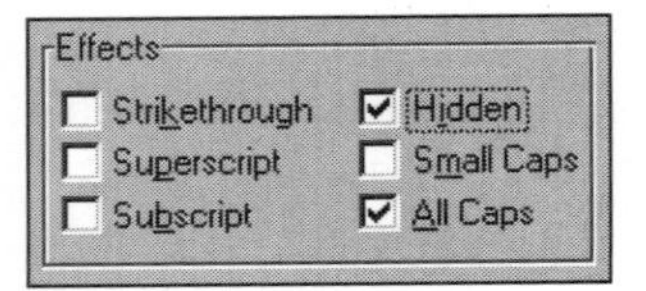

groupware

Acceptable to use to refer to software intended to let a group of users on a network collaborate on a project. Use *messaging system* to describe Microsoft Exchange Server.

grow

Do not use as a transitive verb in the sense of making something larger; use a more specific verb such as *extend* instead.

Correct

If you want to increase your business ...
To extend the selection ...

Incorrect

If you want to grow your business ...
To grow the selection ...

hack, hacker

Do not use *hack* in the sense of writing code or *hacker* to mean an amateur programmer. The words are slang. It may be appropriate to use *hacker* to refer to someone who illegally gains access to a computer system or network with the intent of causing damage.

half inch

Not *half an inch* or *one-half inch.* Hyphenate as an adjective: "a half-inch margin." When space is a concern or the measurement needs to be specific, use *0.5 in.*

handheld (adj)

One word, no hyphen.

Use *Handheld PC* (and the acronym *H/PC*) to refer to the very small computer that runs Windows CE and programs developed for that system. Differentiate from *handheld computer,* which usually refers to specialized computers used in the field in various industries such as transportation.

handle

In technical material, a handle is a pointer to a pointer or a token temporarily assigned to a device or object to identify and provide access to the device. In the latter case, include a space between the word *handle* and the sequential number—for example, "handle 0," "handle 1," "handle 2."

In the interface of various programs, a handle is an element used to move or size an object. Use *move handle* or *sizing handle.* Do not use *size handle, grab handle, little box,* and so on.

handshake (adj, n)

One word. Refers to the connection or signal established between two pieces of hardware (such as a computer and a printer) or communications software (such as the signal to transmit data between two modems). In end-user documentation, briefly define the term at the first occurrence.

Correct

Communicating systems must use the same flow-control (or *handshake*) method. To determine whether the systems use the same handshake method ...

A B C D E F G H I J K L M N O P Q R S T U V W X Y Z

hang

Slang. Avoid; use *stop responding* instead. However, if it's appropriate to your audience, you can clarify your meaning with wording such as "If the program stops responding, or 'hangs,' you may have to restart your computer."

NOTE Sometimes the computer itself stops responding, and sometimes a program does. Be sure any messages refer accurately to the problem.

hard copy (n)

Two words. Okay to use to refer to a paper version of a software document. Avoid using as an adjective.

SEE ALSO **soft copy**

hard disk, hard disk drive

Not *hard drive, internal drive, fixed disk,* or *hard disk system.* Do not hyphenate. See **disk** and **drive**.

hard-coded (adj)

Describes a routine or program that uses embedded constants in place of more general user input. Okay to use in technical documentation, but avoid in end-user material.

hard-of-hearing (adj)

Use the phrase *deaf or hard-of-hearing* to refer to people who have hearing disabilities. If that phrase is too long, use *deaf* only. Do not use *hearing-impaired.*

SEE ALSO **Accessible Documentation, deaf or hard-of-hearing**

hardware

One word.

SEE ALSO **System Requirements**

hardwired (adj)

One word. Describes a functionality that is built into a system's electronic circuitry rather than enabled through software. Avoid this technical term in end-user material.

header

In word-processing and publishing documentation, use instead of *running head* when discussing page layout. However, *running head* is acceptable if needed for clarification or as a keyword or index entry.

In technical documentation, *header* is an acceptable short form of the term *file header,* as in ".rtf header" or "PostScript header." Do not use *header* as a synonym for *header file*, which refers to the file at the beginning of a program that contains definitions of data types and variables used by the program's functions.

Headings and Subheadings

Not *heads* or *headers.*

Headings should convey as much information as possible about the text that follows to help readers locate information quickly.

NOTE Heading style varies among groups, but the guidelines here represent the practice of many and following them helps make sharing files easier. If your group follows a different practice, use it consistently.

Heading style and conventions

These guidelines apply to both online and printed documentation as much as is practicable:

- Avoid referring to the heading in the first sentence following any heading.

  Correct
  Finding Information in Help [Heading]
  When you choose Help from the menu bar, Help commands appear.

  Incorrect
  Finding Information in Help [Heading]
  This is easy to do from the menu bar.

- Use the character formatting that's used in straight text—for example, italic for placeholders. Also, follow the capitalization of any case-sensitive terms, even if it disagrees with the heading style.
- Avoid beginning a heading with an article *(a, an, the).*
- In end-user task-oriented documentation, generally use a present participle (*ing* form) rather than an infinitive (*to*) to begin headings, except for procedure headings. Those should begin with an infinitive phrase. It's acceptable to use "how to" in headings to avoid ambiguity for localization. It's also acceptable in some groups to use the root form of the verb. Use one form consistently.

  Correct
  Running Programs and Managing Files [chapter level]
  Working with Files and Folders [section level]
  Modify a file [topic level]
  To modify a file [procedure heading]

  Incorrect
  To Run Programs and Manage Files
  Work with Files and Folders
  Modifying a file

A B C D E F G H I J K L M N O P Q R S T U V W X Y Z

- For procedure headings, use either the infinitive form or the root form of the verb.

  Correct
  To modify a file
  Modify a file

- For material that isn't task-oriented, you can use descriptive headings when the content requires it.

  Acceptable
  Error Messages and Their Meanings
  Visual Basic Controls
  Accessory Programs

- In printed documentation, avoid headings of more than one line. If a heading must have two lines, try to make the first line longer. For information about permissible line breaks in headings, see **Line Breaks**.
- Headings should refer to singular items unless the plural is obviously called for.

  Correct
  To format a disk
  To open a new document
  Working with Folders

- Use *vs.,* not *versus,* in headings—for example, "Daily vs. Weekly Backups."
- Do not use ampersands (&) in headings unless you are specifically reflecting the interface.
- Use title caps or sentence caps, depending on the level of heading and the design. Book-level headings in Help generally use title caps, but topics use sentence caps.

For rules of capitalization in headings, see **Capitalization**.

Organizational guidelines

Headings help orient users and aid scannability, especially in printed documentation. First-level headings usually apply to the most general material, and subsequent levels deal with more specific topics.

Follow these organizational guidelines for printed documentation:

- Apply the rules for outlining when organizing headings: When dividing a section, try to make the material fall under at least two subheadings. It is permissible to have only one subsection within a section, but only if other methods (such as restructuring the section) will not convey the meaning as well.
- Every heading (except procedure headings) should have text between it and the next heading. However, avoid inserting "filler" text just to adhere to this rule. Intervening text that seems perfunctory or lacking in content is a good indication that restructuring is in order.
- Try to keep headings grammatically parallel within a chapter, section, or other unit, especially those at the same level.

For information about page breaks with headings, see **Page Breaks**.

SEE ALSO **Capitalization, Line Breaks, Lists, Page Breaks, Tables**

hearing-impaired

Do not use; use *deaf or hard-of-hearing* instead.

SEE ALSO **Accessible Documentation**

Help

In general, avoid *online Help;* just use *Help*. However, *online Help, definition Help, context-sensitive Help,* and *online Help files* are acceptable when necessary to describe the Help system itself or to explain how to develop a Help system.

he/she

Do not use. For more information about gender-neutral pronouns, see **Bias-Free Communication**.

hexadecimal (adj)

Do not abbreviate as *hex.* Use *h* or *0x* when abbreviating a number, as in "Interrupt 21h" or "addresses greater than 0xFFFE." Do not insert a space between the number and *h,* and use uppercase for alphabetical characters displayed in hexadecimal numbers.

hierarchical menu

Avoid; use *submenu* instead.

SEE ALSO **submenu**

higher

Do not use to indicate product version numbers; use *later* instead, as in "Microsoft Windows NT version 4.0 or later."

Do use to refer to more powerful hardware, as in "a 386 or higher."

high-level (adj)

Note hyphen. Use the term carefully—a *high-level language,* for example, means a machine-independent language such as Fortran, Basic, or C.

high-level-language compiler

Note both hyphens. Refers to a compiler for a high-level language.

A B C D E F G H I J K L M N O P Q R S T U V W X Y Z

highlight

In general, avoid using *highlight,* unless you are specifically referring to the highlighter feature in some products that users can apply to emphasize selections. Use *select* instead.

Correct

Drag the pointer to select the text you want to format.

Incorrect

Drag the pointer to highlight the text you want to format.

Refer to selected material as *the selection,* not *the highlight.*

Correct

To extend the selection, press F6.

Incorrect

To extend the highlight, press F6.

When it's necessary to be graphically descriptive, you can use *highlight* as a verb to tell the user to select text in a word-processing document, a range of cells in a spreadsheet, or fields and records in a database list view, for example. Likewise, you can use *highlight* to describe the appearance of reverse video. When using *highlight* as a verb in a procedure, include *select* in your procedure so users won't be confused when they use other Microsoft products.

Correct

Highlight the paragraph to select it.
Highlight to select the range of cells you want to copy.
Programmers use reverse video to highlight special items on the screen.

high-quality (adj)

Note hyphen. Do not use *quality* alone as an adjective, only as a noun.

Correct

Use AutoFormat to create high-quality publications easily.

Incorrect

Use AutoFormat to create quality publications easily.

high-resolution (adj)

Refers to a high-quality screen display (generally 640 × 480 pixels or better) or print output (300 dots per inch or better). Note hyphen. Do not abbreviate as *hi-res.*

Hints

Do not use *hint* as a heading for a type of note; use *tip* instead.

SEE ALSO **Notes and Tips**

hi-res

Do not use. See **high-resolution**.

hit

Acceptable to use to refer to the number of times a Web site has been accessed, but note that it's not really an accurate reflection of the number of visits to a site. For example, one page might score a number of hits in one visit, if the user clicks graphics and other internal links. "Page request" is technically the more accurate term.

Do not use to refer to keystrokes; use *press* instead, as in "press ENTER."

home directory

Do not use. Use *root directory* instead to refer to the starting point in a hierarchical file structure. In MS-DOS, the root directory is indicated by a backslash (\).

home page

Refers to the main page of any Web site, as determined by the owner or creator of the site. One Web site can have many home pages. For example, the Microsoft Web site, http://www.microsoft.com, has a home page, but other sites within the Microsoft site have their own home pages.

It also refers to the page the user sets as the first page to view when connecting to the Internet. Do not use *start page*.

Use lowercase to refer to the home page unless you are referring to the command.

host name

Two words.

hot key

Use only to refer to a key or key combination that causes some function to occur in the computer, no matter what else is currently running. A hot key is commonly used to activate a memory-resident program. Do not use to refer to a shortcut key or an access key. Avoid in end-user documentation.

SEE ALSO **Key Names**

A B C D E F G H I J K L M N O P Q R S T U V W X Y Z

hot link

Do not use. It's jargon for a connection between programs that enables information in related databases or files to be updated when information in another database or file is changed. Likewise, do not use to refer to a **hyperlink**.

hot spot

Do not use to refer to a hyperlink. Use in technical documentation to refer to the specific portion (the pixel) on the pointer that defines the exact location to which a user is pointing. Avoid using otherwise except in references to Hotspot Editor, where it's one word.

SEE ALSO **Hyperlinks**

hover, mouse over

Avoid these terms, especially in end-user documentation, to refer to the action of briefly resting the mouse pointer on a button, link, and so on to see a definition or description. Instead, use *rest on, pause on,* or a similar word or phrase.

Do not use *mouse over* as a verb phrase. To describe the action of moving the mouse to a button, use a phrase such as "move the pointer over the button."

how-to vs. how to

Hyphenate as an adjective (as in "how-to books"), but use two words as an adverb plus infinitive (as in "This is how to format your disk").

In headings and titles, do not capitalize the *t,* as in "How-tos, Tips, and Tricks" or "How to Format Your Disk."

HTML

Abbreviation for "Hypertext Markup Language." Spell out as shown on first mention.

HTML is the system of marking a document so it can be published on the World Wide Web and viewed with a browser. Its most distinguishing feature is the hypertext, which means pages can be linked to anything else on the Internet.

Microsoft design guidelines conform to the **W3C** guidelines that allow the pages to be viewed successfully in most browsers. For more information about the W3C, see //www.w3.org.

HTTP

Abbreviation for "Hypertext Transfer Protocol," the Internet protocol that delivers information over the World Wide Web. The protocol appears as the first element in the URL: "http://…." Use lowercase in the URL.

It is acceptable to eliminate "http" from a Web address if you are sure your audience will understand the context. If you are using another protocol such as FTP in an Internet address, however, you must use it.

In general, you do not have to spell out the meaning of the acronym at first mention, unless you are discussing protocols or URLs or to clarify for your audience.

SEE ALSO **Protocols**

Humor

Avoid humor in documentation; it can confuse or even offend users. A jocular tone may sound patronizing to some users, and others may not understand the joke. Humor causes problems in localization, as well. For example, puns can seldom be translated.

Also, in some cultures certain relationships are customarily handled with deference and respect and should not be the subject of jokes, so poking fun at a boss or an older person may be offensive.

Hyperlinks

A hyperlink is the text or graphic that users click to go to a file, a location in a file, an Internet or intranet site, page, or location, and so on. Hyperlinks can also lead to Gopher, telnet, newsgroup, and FTP sites. Hyperlinks usually appear underlined and in color, but sometimes the only indication is that the pointer changes to a hand.

Use *go to* to describe the process of going to another page, and use *create* to describe writing the HTML code that forms the hyperlink. Do not use *hot spot, hot link,* or *shortcut* to refer to a hyperlink. *Link* is acceptable, however.

You can use *followed hyperlink* or *followed link* in technical material to refer to a destination that the user has already visited. Write around in end-user documents.

Correct

Click the hyperlink to go to another Web page.
You can create a link to almost any Web site.

When indicating hyperlinks in Web pages, use the title or a description of the new Web page as the link, rather than a phrase such as "click here." Also, provide short but descriptive **alt text** for graphical links.

Correct

Visit the Editorial Web page for up-to-date style information.

[picture of dinosaur bones]

Incorrect

Click here for up-to-date style information.

[picture]

Hyphens, Hyphenation

Your project style sheet and this guide are the primary sources for hyphenation of product and computer-related terms. However, rules of hyphenation are not always easily applied. If you are in

A B C D E F G H I J K L M N O P Q R S T U V W X Y Z

doubt about whether to hyphenate a modifier, note that the trend is toward less hyphenation. If there's no possibility of confusion, avoid hyphenation. Be sure to note decisions about ambiguous terms on your project style sheet.

For information about hyphenation of common words, see *American Heritage Dictionary* and *The Chicago Manual of Style*. For information about acceptable hyphenation in line endings, see **Line Breaks**. For information about hyphenating with prefixes, see **Prefixes**.

Observe these rules when hyphenating modifiers:

- Hyphenate two or more words that precede and modify a noun as a unit if confusion might result.

  Correct

built-in drive	lower-left corner
high-level language	high-level-language compiler
read-only memory	floating-point decimal
line-by-line scrolling	memory-resident program
scrolling line by line [adverb]	

- Hyphenate two words that precede and modify a noun as a unit if one of the words is a past or present participle.

  Correct
  copy-protected disk
  free-moving graphics

- Hyphenate two words that precede and modify a noun as a unit if the two modifiers are a number or single letter and a noun or participle.

  Correct
  80-column text card
  eight-sided polygon
  8-point font
  16-bit bus
  I-beam insertion point

- Avoid suspended compound adjectives. (A compound word with a suspended hyphen does not include the second part of the compound, such as "first-" in "first- and second-generation computers.") If you must use them to save space, include a hyphen with both the first and second adjectives of the compound. Avoid forming suspended compounds from one-word adjectives that are not hyphenated.

  Correct
  Microsoft Project accepts any combination of uppercase and lowercase letters in a password. [preferable]
  Click the upper- or lower-right corner. [acceptable]

  Incorrect
  Microsoft Project accepts any combination of upper- and lowercase letters in a password.

- Do not hyphenate predicate adjectives (adjectives that complement the subject of a sentence and follow the verb) unless this guide specifically recommends it.

  Correct
  Microsoft Exchange Server is an enterprise-wide messaging system.
  Microsoft Exchange Server controls complicated messaging enterprise wide.
  Many viruses are memory-resident.
  This type of Help is context-sensitive.

- Hyphenate compound numerals and fractions.

  Correct
  his forty-first birthday
  one-third of the page
  three sixty-fourths

- Do not put a hyphen between an adverb ending in *ly* and the verb or adjective it modifies.

  Correct
  Most Internet browsers have a highly graphical interface.

  Incorrect
  Most Internet browsers have a highly-graphical interface.

- Use an en dash (–) instead of a hyphen in a compound adjective in which at least one of the elements is an open compound (such as *Windows NT*) or when two or more of the elements are made up of hyphenated compounds (a rare occurrence).

  Correct
  Windows 98–compatible products
  Some programs have dialog box–type options for frequently used operations.

  Incorrect
  MS-DOS–compatible products

- Do not use a hyphen in key combinations; use a plus sign instead, as in "ALT+O."

SEE ALSO **Capitalization, Em Dash, En Dash, Key Combinations, Line Breaks, Numbers**

I-beam

Note capitalization and hyphenation. Avoid specific references to the I-beam pointer (instead, refer simply to *the pointer*), except when necessary to describe how the pointer's shape affects its function—for example, "when you click text, the pointer looks like an I-beam."

icon

Use only to describe a graphic representation of an object that a user can select and open, such as a drive, disk, folder, document, or program.

Icons

When referring to a program icon, use bold for the icon name: "Click the **Word** icon." Within programs, do not use *icon* for graphical dialog box options or options that appear on ribbons, toolbars, toolboxes, or other areas of a window.

In general, use the most descriptive term available, such as *button, box, check box, tool,* and so on. If an option has no visual properties except its graphic nature, use *symbol,* as in "warning symbol."

iconize

Do not use; instead, use *shrink to an icon* or *minimize.*

Identifiers

Do not capitalize identifiers such as *menu, command, button,* or *box.*

Correct

Cancel button
File menu
Print dialog box

i.e.

Means *id est.* Do not use; use *that is* instead.

if vs. when vs. whether

To avoid ambiguity, use *if* for uncertainties or conditionals, use *when* for situations requiring preparation or the passage of time, and use *whether* for one or more alternative possibilities or situations.

Correct

The printer might insert stray characters if the wrong font cartridge is selected.
If your document will not print ...
To find out whether TrueType fonts are available or ...
When you are ready to print your document ...

Incorrect

To find out if TrueType fonts are available ...
When your document will not print ...

imbed

Do not use; use *embed* instead.

impact (n)

Do not use as a verb—it's jargon. Use *affect* or another synonym instead.

Correct

Sending inappropriate e-mail can affect your career adversely.

Incorrect

Sending inappropriate e-mail can impact your career adversely.

Important Note

A note labeled "Important" that provides information essential to the completion of a task. Users can disregard information in a note (labeled "Note" or something similar) and still complete the task, but they should not disregard an important note.

SEE ALSO **Notes and Tips**

in, into

In indicates within the limits, bounds, or area of or from the outside to a point within. *Into* generally implies moving to the inside or interior of.

A B C D E F G H I J K L M N O P Q R S T U V W X Y Z

Correct

A word is in a paragraph, but you move the text into the document.
Data is in a cell on a worksheet.
You edit the cell contents in the formula bar.
A file name is in a list box.
A workstation is in a domain, but resources are on servers.
You open multiple windows in a document.
You insert the disk into the disk drive.
You run programs with, on, or under an operating system, not in them.

in order to

A verbose phrase that is usually unnecessary. Use just *to* instead.

inactive, inactive state

Not *not current.*

in-bound

Avoid in the sense of messages being delivered. Use *incoming* instead.

incent (v)

Marketing jargon; do not use. Use a more specific term instead.

Correct

This pricing should encourage users to buy the new version.

Incorrect

This pricing should incent users to buy the new version.

incoming, outgoing

Use to refer to e-mail messages that are being downloaded or being sent. Avoid *in-bound* and *out-bound.*

increment (n, v)

In programming material, restrict the use to mean "increase by one." In more general material, it's okay to use to mean "grow by regular consecutive additions."

indent, indentation

Use *indent,* not *indentation,* to refer to a single instance of indentation—for example, *hanging indent, nested indent, negative indent* (do not use *outdent*), *positive indent.* Do not use *indention.* Use *indentation* to refer to the general concept.

indenting, outdenting

Do not use "indenting or outdenting into the margin." Instead use "extending text into the margin" or "indenting to the previous tab stop."

independent content provider

Okay to use the abbreviation "ICP" after spelling out. Refers to a business or organization that supplies information to an online information service such as MSN or America Online.

index (s), indexes (pl)

Use *indices* only to refer to mathematical expressions.

Indexing

Indexes and keywords are critical information-access tools. When you develop index entries, consider the tasks the user will want to accomplish, previous versions of the Microsoft product that may have used different terms, and the terminology of similar products the user might be familiar with. These principles are the same for both printed and online indexes.

This topic describes some indexing concepts, but it focuses more on mechanical issues such as alphabetizing, style, and cross-references and pertains primarily to printed indexes. For developing search keywords and online index entries, see **Keywords and Online Index Entries**.

Creating entries

Users of documentation depend on indexes as their primary way to find the information they need, so an index must be complete, thorough, and accurate. Although the number of indexed terms per page will vary depending on the subject and complexity of the book, a rule of thumb is that a two-column index should be about 4 to 8 percent of the total number of pages in the book.

When you create new main entries, place the important word first. Depending on the kind of material, that word should probably be a noun (*commands, addresses, graphs*), but it can be a gerund (*copying, selecting*). Do not use nonessential words as the first in an entry and do not use vague gerunds such as "using" or "creating."

Try to think like a user. A user who wants to delete paragraphs will probably look for the information under "paragraphs" and "deleting," possibly under "Delete command," but most likely not under "using Delete."

Invert entries whenever possible. For the previous example, you would include an entry for "paragraphs, deleting" and one for "deleting paragraphs." Other examples include items such as "arguments, command line" and "command-line arguments." Page numbers for inverted entries should match exactly.

Likewise, if you use synonyms to help the user find information in more than one place in the index, modifiers, subentries, and page numbers should match.

A B C D E F G H I J K L M N O P Q R S T U V W X Y Z

Creating subentries

For every main entry you create, make a subentry. You may have to delete those subentries later, but that's easier than having to add them.

Consider as subentries these generic terms, especially when a topic is covered in various places in the book: defined (to refer to a term), described (to refer to an action), introduction, overview.

Avoid the use of prepositions to begin subentries. However, sometimes prepositions can clarify relationships.

Do not repeat a main term in the subentry.

Correct

pointers
    far
    function call

Incorrect

pointers
    far pointers
    function call

Avoid using more than five page references after an entry. Subentries give more direction to the user. Do not, however, use only one subentry; there must be two or more.

Correct

paragraphs, deleting 72

Correct

paragraphs
    deleting 72
    formatting 79, 100

Incorrect

paragraphs
    deleting 72

Incorrect

paragraphs 72, 75, 79, 100, 103, 157

If possible, use only one level of subentries, but never more than two. Localized indexes can become unreadable with two levels of subentries because entry words are often lengthy and must be broken. If your group decides to use two levels of subentries, you should inform the localization team as early as possible.

Correct (one level of subentries)

paragraph formatting 75
- characters and words 63
- using styles 97

paragraphs, deleting 72

Correct (two levels of subentries)

paragraphs
- deleting 72
- formatting 75
  - characters and words 63
  - using styles 97

If a main entry is followed by subentries, do not leave the main entry as a widow at the bottom of a column. Also, if a list of subentries is long and will run over to a second column, repeat the main entry, followed by the word *continued,* which is lowercase, in parentheses, and italic (including parentheses), at the beginning of the second column. If the column break occurs between two second-level subentries, repeat the main entry, followed by *continued,* and the subentry, also followed by *continued.* The word *continued* is lowercase, in parentheses, and italic (including the parentheses). Avoid leaving only one subentry before or after column breaks.

Correct (main entry continued)

paragraphs
- deleting 101

paragraphs *(continued)*
- formatting 87–96
- indenting 98

Correct (subentry continued)

paragraphs
- deleting 101
- formatting 87–96

paragraphs *(continued)*
- formatting *(continued)*
  - characters and words 63
  - using styles 97

Incorrect

paragraphs
- deleting 101

- formatting 87–96
- indenting 98

Incorrect

paragraphs

paragraphs *(continued)*
- deleting 101
- formatting 87–96
- indenting 98

A B C D E F G H I J K L M N O P Q R S T U V W X Y Z

Page references

Separate multiple page references with commas. Separate page ranges with en dashes surrounded by a thin space; however, you can use hyphens to indicate page ranges if you need to conserve space. Do not abbreviate page references.

If possible, avoid long multiple page references listing consecutive pages. A page range might better represent the topic. Likewise, avoid chapter or section length page ranges if the topic is clearly shown in the table of contents. Users prefer to be able to find more specific information.

Correct

paragraphs 24, 47, 126–130
    deleting 72–76
    formatting 87

Incorrect

paragraphs 24, 47, 126, 127, 128, 129, 130
    deleting 72, 73, 74, 75, 76
    formatting 87

Style and formatting

Use the index style in the same Microsoft design template you used for your document. The font should be the same as that in the book, but in a smaller point size.

In general, do not use special character formats such as bold, monospace, or italic for entries. Use italic for cross-references (*See* and *See also* references).

Capitalization

Because many groups use the same source for both printed and online documentation, all lowercase is recommended for all index entries except those words that require capitalization and *See* and *See also* references.

If a group knows that the index will not appear online or be shared with another group's documentation, that group can capitalize entries.

Plural vs. singular

Use the plural form of all main entries that are nouns, except where it's awkward or illogical to do so. The following table shows correct use of both plural and singular.

Correct use of plural	*Correct use of singular*
borders	File command
files	e-mail
headers	ruler
paragraphs	window

Prepositions and articles

Limit the use of prepositions and articles. Use them only when they are necessary for clarity or sense. In general, do not use articles unless required for clarity.

Correct

child windows
open, list of 128
opening 132, 137
reading from 140
writing to 140
structures, in programming for Windows 200

Verbs

Use a gerund rather than the infinitive or the base present tense form for entries about actions, processes, or procedures.

Correct

selecting
drawing objects 22
text 147
shapes
drawing 37
fitting around text 140
fitting text into 131
substituting text 255

Incorrect

select
art 255
text 147
shapes
to draw 37
to fit around text 140

Versus vs. vs.

Use the abbreviation *vs.* (including the period) in index entries.

Correct

Voice, active vs. passive 98

Cross-references in indexes

An index can have the following types of cross-references:

See
See also
See specific [name of item]
See also specific [name of item]
See herein

A B C D E F G H I J K L M N O P Q R S T U V W X Y Z

Format the *See, See also,* and *See herein* phrases in italic, and capitalize *See* to avoid confusion with the actual entries. Use lowercase for the name of the entry referred to.

Place *See* cross-references on the same line as the entry, separated by two spaces. Place *See also* references on a separate line and sort them as the first subentry. (Optionally, if the main entry has no other subentries, you can place a *See also* reference on the same line. See the Pontoons entry in the following example.)

Do not use page numbers with cross-references. Alphabetize multiple topics following one cross-reference and separate them with semicolons.

Correct

- airplanes *See* planes
- airports *See specific airport*
- floatplanes 101–105
- planes
  - *See also specific plane*
  - rudders
    - control 66–67
    - types 61
  - steering
    - *See also* rudders
    - guidelines 45
    - taxiing *See herein* takeoff
  - takeoff
    - control tower 19
    - steering 22, 25, 27
- pontoons 98 *See also* floatplanes
- seaplanes
  - *See also* aeronautics; floatplanes; pontoons
  - rudders
    - controls 66–67
    - types 61
- steering *See* rudders
- water
  - *See also* seaplanes
  - taking off on 18

Order of entries

Special characters appear at the beginning of the index, followed by numeric entries, sorted in ascending order. Alphabetical entries then follow. Separate the categories with headings if there are many of them; if there are only a few, no special separation is necessary. Use the heading *Symbols* for special characters if you use a heading.

Alphabetizing indexes

Microsoft alphabetizes word by word, not letter by letter. That is, words separated by spaces or commas are treated as two words. Alphabetizing stops at the end of a word unless the first word of two or more entries is the same. Then the first letter of the second word determines alphabetical order, and so on. Letter-by-letter alphabetization ignores spaces, treating each entry as one word. Compare the columns in the following table to see the difference. For more information, see *The Chicago Manual of Style.*

Word by word	*Letter by letter*
D key	Delete command
DEL key	deleting
Delete command	DEL key
deleting	D key

Special characters

Index special characters at least twice. List each character by its symbol, followed by its name in parentheses, as the next example shows. Also list each character by name, followed by its symbol in parentheses. You might also want to index some characters under a general category, such as "operators."

Special characters that are not part of a word are sorted in ASCII sort order. The name of the character follows in parentheses. They appear at the beginning of the index, followed by numeric entries.

Correct

% (percent)
& (ampersand)
((opening parenthesis)
) (closing parenthesis)
* (asterisk)
| (pipe)
~ (tilde)

Special characters followed by letters or within a word are ignored in alphabetizing and are usually included in the alphabetical listing. Sometimes, however, you may want to include such entries in both the alphabetical list and in the list of special characters.

Correct

Error
errors, correcting
^p
paragraphs
#VALUE
values

A B C D E F G H I J K L M N O P Q R S T U V W X Y Z

Sorting numbers as numbers

Numeric entries should be placed in ascending order, with entries containing only numbers falling before those containing both numbers and letters. This requires editing to correct the machine sort. Compare these two lists of sorted numerics:

Machine sorted	*Edited*
12-hour clock	80386
2-D chart	80486
24-hour clock	2 macro
80386	2-D chart
80486	3-D chart
1904 date system	12-hour clock
366-day year	24-hour clock
3-D area chart	366-day year
2 macro	1900 date system
1900 date system	1904 date system

Numbers follow the list of special characters.

Sorting spaces and punctuation

Entries that have the same letters but different spacing or punctuation are governed first by the rules of word-by-word alphabetization and then by the rule for sorting special characters by their ASCII order: Spaces alone come first, then spaces following commas. Next come unusual connecting characters, in ASCII order: periods, colons and double colons, underscores, and so on. Apostrophes, hyphens, and slashes are ignored, so those entries will come last, in that order.

Correct

_name changers
name changers
name, changers
NAME.CHANGERS
name:changers
name_changers
.namechangers
namechangers
namechanger's
name-changers
name/changers

initialize

Technical term usually referring to preparing a disk or computer for use or to set a variable to an initial value. Do not use to mean start a program or turn on a computer.

initiate

Do not use to mean start a program; use *start* instead.

inline (adj)

One word, no hyphen. "Inline styles" are used in cascading style sheets to override a style in the style sheet itself. Inline styles are embedded in the tag itself by using the STYLE parameter.

input (n)

Avoid using as a verb; use *type* instead.

Correct

Word moves existing characters to the right as you type new text.

Incorrect

Word moves existing characters to the right as you input new text.

input device

Use only in a general discussion of methods of inserting information. Do not use to describe the mouse, which is only one type of input device.

SEE ALSO **mouse, pointer**

input/output

In general, spell out at first use, and then abbreviate as *I/O.* However, some technical audiences might be completely familiar with the term, in which case it is not necessary to spell out at first use.

input/output control

Acceptable to abbreviate as *I/O control* or *IOCTL* after first occurrence. Use only in technical material.

insertion point

The point at which text or graphics will be inserted when the user begins working with the program. It's usually shown as a blinking line or, in character-based applications, a blinking rectangle. Use instead of *cursor* except in character-based applications, where *cursor* is acceptable. Always use the article *the,* as in "the insertion point."

inside

Use instead of the colloquial *inside of.*

A B C D E F G H I J K L M N O P Q R S T U V W X Y Z

install

In general, use *install* to refer to adding hardware such as printers or CD-ROM drives to a system and to adding software to a hard disk.

instantiate

Avoid. Use *create an instance of* [a class] instead.

insure

Do not use except to refer to insurance. Use *ensure* to mean "to make certain."

interface (n)

Use as a noun only, as in "user interface" and "programming interface." Use *on* as the preposition preceding *interface.* Avoid as a verb; it's jargon. Use *interact* or *communicate* instead.

Correct

It's easy to use the Internet to communicate with various interest groups.
The interface is so intuitive that even first-time users get up to speed quickly.
The color can be adjusted on the interface.

Incorrect

It's easy to use the Internet to interface with various interest groups.
The color can be adjusted in the interface.

In COM-based technologies, an interface is a collection of related public functions called *methods* that provide access to a COM object. The set of interfaces on (note preposition) a COM object composes a contract that specifies how programs and other objects can interact with the COM object.

SEE ALSO **COM, ActiveX, and OLE Terminology**

International Considerations

Localized versions of Microsoft products and documentation follow the U.S. version as closely as possible in wording, organization, art, layout, and so on. Making the U.S. source material as clear, correct, and internationally relevant as possible is essential to smooth localization.

Try to think "international" from the earliest product stages on, and include relevant localization staff in planning and the documentation review.

Writing for translation

Because most documentation is localized by third-party vendors in the target country, it's very important for the English source material to be highly translatable. Simple mistakes, ambiguity, or inappropriate use of terminology can surface as queries or errors in translation, requiring revision in more than 20 language versions of the same product. Clear, concise, and grammatically correct English writing results in shorter review cycles for translated material and reduced costs.

Use correct terminology and consistent wording

- Define and use the appropriate terminology, according to the software and this guide.

  Localizers and translators work with bilingual glossaries that contain technical terms and phrases found in the product interface and documentation. Using correct and consistent terminology saves translation time (because glossaries can be shorter and simpler), allows some automatic translation, and allows vendors to use multiple translators on a project. For example, commands are *clicked,* not *selected.* (For more information, see **Procedures**.)

  Correct
  ... when you click Create ...

  Incorrect
  ... when you select the Create command ...

- Use only one term to name one concept.

  Using the word *tab* for five different concepts, for example, forces the translator to choose the correct translation in any given context, which may result in an error. The solution is to use *tab* for dialog box panels only, to use the entire term *(tab stop, tab character,* and so on) in the other cases, and to use an alternative verb (such as *move*) to describe movement of the insertion point using the TAB key.

  Correct
  ... using the TAB key to move through a dialog box
  ... setting a tab stop
  ... moving a tab mark on the ruler
  ... how to show or hide tab characters
  ... the View tab in the Options dialog box

  Incorrect
  ... to tab through a dialog box
  ... setting a tab
  ... moving a tab on the ruler
  ... how to show or hide tabs

  A similar problem is using several terms to express the same concept. For example, the words *create, add,* and *insert* are often used for the same action. The solution is to choose one term for each concept and use it consistently.

- Use both words and phrases consistently. Localization glossaries now include phrases and even entire paragraphs. When all of the following phrases are used in a document instead of just one of them, the localizers are unable to reuse already translated material.

  Lack of consistency
  The following prompt will appear ...
  You will be prompted with ...
  This prompt appears ...

A B C D E F G H I J K L M N O P Q R S T U V W X Y Z

A B C D E F G H I J K L M N O P Q R S T U V W X Y Z

Avoid ambiguity

- In colloquial English, words that might have clarified a sentence are sometimes omitted. This is most often the case of prepositions and verbs used with gerunds. Because gerunds are not found in most languages, they represent a special challenge for the translator; do not omit the accompanying words that are essential to the meaning.

  Correct
  You can change the Admin.tpl and the Admin.inf files that are using the Template utility.
  You can change the Admin.tpl and the Admin.inf files by using the Template utility.
  This prevents you from accidentally losing your work if your computer runs out of memory.

  Incorrect
  You can change the Admin.tpl and the Admin.inf files using the Template utility.
  This prevents you from accidentally losing your work by your computer's running out of memory.

- Avoid overmodified nouns, or noun stacks, which are especially difficult to translate when several different combinations could make sense, as in the first Incorrect example.

  Correct
  For new users, set the parameters to the default printer configuration.
  Enter the maximum length of time, in number of days, that you want to keep the address lists updated by automatic synchronization. Then press ENTER.

  Incorrect
  Set default printer configuration parameters for new users.
  Enter the maximum length of time that you want to keep the automatic synchronization address list updates in days and press ENTER.

  Note that in some cases, if the context is clear, some nouns can be eliminated. "Automatic synchronization" may be unnecessary in the preceding examples, especially because the focus is on updated address lists.

- Be specific in defining a role versus a person.

  Correct
  To create another administrator account, click New on the File menu.
  To set privileges for another administrator, click New on the File menu.

  Incorrect
  To create another administrator, click New on the File menu.

Avoid long and complex sentences

A complex sentence requires the translator first to understand the content; then choose the main clause, which the translation should begin with; and then translate it. When conveying complex ideas, put main ideas first and break the material into small units. If necessary, clarify by reformatting.

Correct

The Dispatch program is used for directory synchronization. It runs the programs that transfer both the local address updates from the requestors to the directory server and the global address updates back to the requestors for processing.

Incorrect

Directory synchronization uses the Dispatch program to run the programs that transfer the local address updates from the requestors to the directory server and to transfer the global address updates back to the requestors for processing.

Easy to follow

However, if the newly attached data file contains field names different from those inserted in the main document, you must do one of the following:

- Change the field names in the header record to match the field names in the main document.
- Replace the field names in the main document.

Difficult to follow

However, if the newly attached data file contains field names different from those inserted in the main document, you must either change the field names in the header record to match the field names in the main document or replace the field names in the main document.

Use commas correctly

Set off introductory phrases and clauses with commas.

Correct

To manage your files, click the Start button ...
Using Control Panel, you can change the way Windows looks and works.

Use verbs as verbs and nouns as nouns

Unlike many other languages, many English words can be used as both nouns and verbs, and nouns can be turned into verbs and vice versa. To avoid confusion or ambiguity, use standard and consistent terminology for computer actions. Try to find unambiguous synonyms for other words.

Correct

If you edited your document ...
... information that you want to refer to

Incorrect

If you made edits to your document ...
... information that you want to reference

Follow standard English word order

When two or more correct arrangements are possible, choose the order that will create the least ambiguity. Generally, this is subject-verb-object, with modifiers before or immediately following what they modify.

Standard word order

Your primary concern may be to update the address lists.

A B C D E F G H I J K L M N O P Q R S T U V W X Y Z

Inverted word order

To update the address lists may be your primary concern.

Avoid idiomatic or colloquial expressions

Idioms and colloquial expressions are usually untranslatable. Many have no counterparts in other languages, or their use is considered inappropriate for the intended audience. Often, translators will substitute a more businesslike phrase for an idiomatic expression or leave the expression out altogether. In any case, the translator may have to take time to understand the meaning.

Correct

Now you have an instant design.
These simple steps complete the task.

Colloquial

Now how's that for instant design!
That's all there is to it.

Avoid wordiness and jargon

Succinct sentences are the easiest to translate. Jargon and wordiness blur the meaning and are hard to convert into another language and syntax. Also, because translators are paid by the word, wordiness is costly.

Concise

You've concluded Lesson 4 and learned to use the drawing tools and text editing buttons. You can continue now, or leave the tutorial and return to Lesson 5 later. For more information, see "Saving and Quitting."

Wordy

Now is probably a good time to save your presentation again. If you want to take a break before the next lesson, go ahead and exit PowerPoint. Refer to the steps noted earlier in this chapter if you need more information.

You've concluded the fourth lesson and learned the basics of using the drawing tools and the text editing buttons. If you're ready to do more, just keep going. Or you can take a break and come back to Lesson 5 another time.

Samples, scenarios, and art

Unless samples, scenarios, and art are planned with cultural diversity in mind, artwork must be created from scratch for localized product versions. Because samples and visual material are never re-created for the English product versions sold outside the United States, this material should be universally appealing as well as culturally appropriate.

Symbols and metaphors

Some symbols or images pose problems for international versions. For example, in localized products for some countries, product wizards are referred to as *assistants,* resulting in the loss of the symbolic value and figurative meaning of the original term. Because no semantic equivalent could be used, the supporting sorcery-related images, such as magic wand buttons or starry tutorial backgrounds, couldn't be used.

Some symbols common in the United States may not be recognized in other countries. A U.S. style mailbox, for instance, may not mean the same thing in other places.

Colors are often described in floral terms, but the translator may miss the meaning, as happened with an example involving the color periwinkle. Although in English this word describes the light blue of the periwinkle blossom, its translation into German is *Immergrün* ("evergreen") because there the plant is prized for its dark green leaves.

Scenarios

The goal for U.S. teams is to use interesting yet internationally accessible scenarios. Inappropriate scenarios require re-creating art and rewriting text. For example, scenarios involving gourmet dog food or varieties of birdbaths are problematic because these luxury consumer goods are unknown in many countries where English or localized product versions are sold.

Even if a chosen scenario works for different cultures, however, research is often necessary to use it. Although mountain climbing is an internationally accessible subject, some research may be required for the translator to do the job. What makes climbing an acceptable subject is that relatively little research (say, buying a magazine on the subject) will give the translator the necessary information.

Should everything be "internationalized"?

Although examples specific to the United States should be kept to a minimum, they are sometimes necessary. For example, a payroll spreadsheet might be the best way to illustrate the Microsoft Excel features, and localized versions would have to be created from scratch for every language version. In cases like these, localizers should be notified early so these documents can be created ahead of time and not during the localization cycle.

Also, it is sometimes necessary for writers to use words or phrases symbolically. For example, writers in Windows NT used the word *tombstone* to mean the oldest entry in a database. After they told localizers how they were using the term, localizers were able to translate it without a problem.

Text expansion

Text expands in length during translation. In most cases, text increases by 10 or 15 percent; in some languages, localized text can be as much as 30–35 percent longer than the source text. These figures are for normal text paragraphs, where expansion can usually be accommodated (for example, in the last short line of each paragraph, or in the white space at the end of a chapter). Short text pieces, such as callouts, can expand much more drastically.

The following sample translations (English to German) demonstrate the different rates of expansion for larger and smaller text elements.

English	*German*
50 characters To see more ports, use the UP and DOWN ARROW keys.	**92 characters (84 percent increase)** Drücken Sie die NACH-OBEN-TASTE oder die NACH-UNTEN-TASTE, um weitere Anschlüsse anzuzeigen.
63 characters Drag these markers ... ... to create a hanging indent in the text.	**101 characters (60 percent increase)** Ziehen Sie diese Markierungszeichen, ... um im Text einen negativen Erstzeileneinzug zu erstellen.

Consider the expansion factor when planning manual size and document structure as well as when writing. For instance, if the margin contains too much text, the translated element could extend beyond the body text it relates to. Worse, it might extend into the next marginal text element. Callouts should not contain entire procedures because their translated counterparts often look cluttered and unprofessional. In these cases, reorganize or rewrite to keep margin text or callouts short.

Internet, intranet

The term *the Internet* (capitalized) refers to the worldwide collection of networks that use the TCP/IP protocols to communicate with each other. The Internet offers a number of tools, including e-mail, the World Wide Web, and other communication services.

The term *internet* (lowercase) refers to any large network made up of a number of smaller networks. In general, avoid use of the lowercase term or define it so that it's not confused with the Internet.

An *intranet* (lowercase) is a communications network based on the same technology as the World Wide Web that's available only to certain people, such as the employees of a company.

For more information, see the *Microsoft Press Computer Dictionary.*

Internet Explorer

Always use *Microsoft Internet Explorer* in first-level topic headings in Help, at first mention on a Web page, and at first mention in a chapter. Then use *Internet Explorer*. Never use *IE.*

Internet service provider

Okay to use the abbreviation *ISP* after spelling out. Refers to an organization that provides access to the Internet. ISPs also usually provide services such as e-mail and newsgroups, and may contain proprietary content. Connection to the ISP is usually through a modem and phone line or over a dedicated line such an ISDN line. Many smaller ISPs serve as a host for independent Web sites.

Some major ISPs are MSN and America Online. These larger services are sometimes considered "online services" rather than ISPs because they maintain their own databases and forums, but the distinction is minor.

Interrupt

When discussing MS DOS interrupts, spell out and capitalize the word *interrupt* and use a lowercase *h,* as in "Interrupt 21h."

inverse video

Avoid; use *reverse video* instead. Use *highlighted* to refer to the appearance.

SEE ALSO **highlight**

invoke

Do not use in end-user documentation in the sense of starting or running a program; it's jargon. Acceptable in technical documentation to refer to a function, process, and similar elements.

IP address

The numeric Internet Protocol address assigned by the Network Information Center (NIC) that uniquely identifies each computer on the network that uses TCP/IP. The IP address is a 32-bit identifier made up of four groups of numbers, each separated by a period, such as 123.432.154.12. It's sometimes called the "dotted quad," but don't use that term in documentation.

issue

Avoid using as a verb; try to use a more specific verb instead. Do not use to refer to commands in end-user documentation.

Correct

Windows 95 displays an error message.
Click Save As to save a file under a new name.

Incorrect

Windows 95 issues an error message.
Issue the Save As command to save a file under a new name.

italic (adj)

Not *italics* or *italicized.*

SEE ALSO **Document Conventions, font and font style**

its vs. it's

Proofread your work to be sure you've used the correct word. *Its* is the possessive form; *it's* is the contraction meaning "it is."

Correct

It's easy to take advantage of many new features in Office.
The easy connection to other systems is just one of its many advantages.

Jargon

Jargon is a neutral term—it refers only to the technical language used by some particular profession or other group. It can be acceptable in the right context for a particular audience, where it serves as a shortcut to understanding concepts for those who understand the term.

Technical jargon is often acceptable in documentation for programmers and other technical audiences where you can assume a certain background or level of expertise. Marketing jargon and buzzwords generally are not acceptable. You can find a list of jargon in the index. Some of the terms are acceptable, especially in certain circumstances, and some are not.

Jargon is not acceptable if:

- A more familiar term could easily be used.
- It obscures rather than clarifies meaning, as some new terms or terms familiar only to a small segment of the audience may (*glyph,* for example).
- It's not specific to computer software, networks, operating systems, and the like. That is, avoid marketing and journalistic jargon (*leverage,* for example) in documentation.

A technical term that's specific to a product should be defined and then used without apology.

Testing for jargon

If you're familiar with a term, how can you tell if it's jargon that you should avoid? If the term isn't listed either in this book or on your project style sheet, check the following list:

- If in doubt, it's probably jargon.
- If an editor or reviewer questions the use of a term, it may be jargon.
- If it's used in other documentation, either Microsoft's or another company's, it's probably acceptable.
- If it's used in technical articles in newspapers such as the *Wall Street Journal* or the *New York Times* or mass market magazines such as *PC Week,* it's probably acceptable.

Java, JScript, JavaScript

Java is an object-oriented programming language developed by the Sun Corporation. Microsoft's implementation of Java is J++. Because it is platform-independent, it is especially useful for programming **applets** for the World Wide Web.

Do not confuse Java or J++ with JScript or JavaScript. JScript, a Microsoft product, is an object-based scripting language distantly and loosely related to Java. JavaScript is a similar Netscape product, so refer to JScript in Microsoft documentation.

join (adj, v, n)

Do not use to mean "embed." *Join,* in database terminology, refers to a relationship or association between fields in different tables and should be reserved for that meaning in documentation for database and related products.

Correct

If you join numeric fields that do not have matching FieldSize property settings, Microsoft Access might not find all the matching records when you run the query.
When you add fields from both tables to the query design grid, the default, or inner, join tells the query to check for matching values in the join fields.
To embed one object into another, click Paste on the Edit menu.

Incorrect

To join one object with another, click Paste on the Edit menu.

joystick

One word. Joysticks have *controls* (not options) for controlling movement on the screen.

jump

Do not use as a noun to refer to cross-references to other Help topics or to hyperlinks. Likewise, do not use as a verb to refer to going from one link to another; use *go to* instead.

justify (v), justified (adj)

Avoid except as a synonym for or cross-reference to *align.* Then use only to refer to text that is aligned with both the left and right margins. Do not use *left-justified* or *right-justified;* use *left-aligned* or *right-aligned* instead.

KB

Abbreviation for *kilobyte.* Use the abbreviation only as a measurement with numerals; do not use in straight text without a numeral. Spell out *kilobyte* at first mention if your audience may not be familiar with the abbreviation. Insert a space between *KB* and the numeral.

Correct

360-KB disk
64 KB of memory left

SEE ALSO **kilobyte**

K, K byte, Kbyte

Do not use as abbreviations for *kilobyte.* Use *KB* instead. Do not use *K* to refer to $1000. It's slang. See **KB**.

Kbit

Do not use as an abbreviation for *kilobit;* always spell out.

KBps, Kbps

KBps is the abbreviation for *kilobytes per second. Kbps* is the abbreviation for *kilobits per second.* Use the abbreviations only as a measurement with numerals; do not use in straight text without a numeral. Spell out at first mention.

Kerberos protocol

The Kerberos version 5 authentication protocol is a feature of Windows NT 5.0 Distributed Services architecture. It is often referred to as Kerberos V5. Be sure to differentiate between Kerberos V4 and V5 protocols. Windows NT supports only the Internet standard version of the protocol, which is version 5.

Always use *Kerberos* as an adjective ("Kerberos protocol"), not as a noun ("includes Kerberos"). Always make clear in the first reference to the Kerberos protocol that Windows NT implements version 5.

Correct

Microsoft Windows NT includes support for the Kerberos V5 protocol.
Microsoft security services are compatible with the Kerberos V5 authentication.
Kerberos V5 protocol is a feature of Windows NT 5.0 security.

Incorrect

Microsoft Windows NT includes support for Kerberos.
Microsoft security services are compatible with Kerberos.
Kerberos is a feature of Windows NT 5.0 security.

Key Combinations

Use a plus sign, as in ALT+O, to indicate key combinations such as shortcut keys and access keys. In the first example, the user would press and hold down ALT and then press O.

Correct

ALT+O
CTRL+P

"Shifted" key combinations

To show a key combination that includes a "shifted" key, such as the question mark, add SHIFT to the combination and give the name or symbol of the shifted key, such as ? or $. Using the name of the unshifted key, such as 4 rather than $, could be confusing to users or even wrong; for instance, the ? and / characters are not always shifted keys on every keyboard. Do, however, spell out the names of the plus and minus signs, hyphen, period, and comma.

Correct

CTRL+SHIFT+?
CTRL+SHIFT+*
CTRL+SHIFT+COMMA

Incorrect

CTRL+SHIFT+/
CTRL+?
CTRL+SHIFT+8
CTRL+*

SEE ALSO **Key Names**, **Key Sequences**

Key Names

In general, spell key names as they appear in the following list, whether the name appears in text or in a procedure. Use all caps unless otherwise noted.

> **NOTE** This list applies to Microsoft and IBM-type keyboards unless otherwise noted. For more information about keys on the Microsoft Natural Keyboard, see "Microsoft Natural Keyboard key names," on page 146. Differences with the Macintosh keyboard are noted.

Correct

ALT

ALT GR

Application key [Microsoft Natural Keyboard only]

arrow keys [not *direction keys, directional keys,* or *movement keys*]

BACKSPACE

BREAK

CAPS LOCK

CLEAR

COMMAND [Macintosh keyboard only. Use the bitmap to show this key whenever possible, because the key is not named on the keyboard.]

CONTROL [Macintosh keyboard only. Does not always map to the CTRL key on the PC keyboard. Use correctly.]

CTRL

DEL [Macintosh keyboard only. Use to refer to the forward delete key.]

DELETE [Use to refer to the back delete key on the Macintosh keyboard.]

DOWN ARROW [use *the* and *key* with the arrow keys except in key combinations or key sequences. Always spell out. Do not use graphical arrows.]

END

ENTER [On the Macintosh, use only when functionality requires it.]

ESC [Always use ESC, not ESCAPE or Escape, especially on the Macintosh]

F1–F12

HELP [Macintosh keyboard only. Always use "the HELP key" to avoid confusion with the Help button.]

HOME

INSERT

LEFT ARROW [use *the* and *key* with the arrow keys except in key combinations or key sequences]

NUM LOCK

OPTION [Macintosh keyboard only]

PAGE DOWN

PAGE UP

PAUSE

PRINT SCREEN

RESET

RETURN [Macintosh keyboard only]

RIGHT ARROW [use *the* and *key* with the arrow keys except in key combinations or key sequences]

SCROLL LOCK

SELECT
SHIFT
SPACEBAR [precede with *the* except in procedures, key combinations, or key sequences]
SYS RQ
TAB [use *the* and *key* except in key combinations or key sequences]
UP ARROW [use *the* and *key* with the arrow keys except in key combinations or key sequences]

Windows logo key [Microsoft Natural Keyboard only]

Spell key names that do not appear in this list as they appear on the keyboard.

When telling a user to "press" a key, format the key name in all caps. When telling a user to "type" a key, use lowercase bold, unless an uppercase letter is required.

Correct

Press Y.
Type **y**.

> **NOTE** Format punctuation according to intended use. If the user must type the punctuation, use bold. If not, use roman.

At first mention, you can use *the* and *key* with the key name if necessary for clarity—for example, "the F1 key." At all subsequent references, refer to the key only by its name—for example, "press F1."

For the arrow keys and the TAB key, list only the key name in key combinations without *the* and *key.*

Correct

To move the insertion point, use the LEFT ARROW key.
To extend the selection, press SHIFT+LEFT ARROW.

Special character names

Because these keys could be confused with an action (such as +) or be difficult to see, always spell out the following special character names: PLUS SIGN, MINUS SIGN, HYPHEN, PERIOD, and COMMA.

Correct

SHIFT+PLUS SIGN
Press ALT, HYPHEN, C
Press COMMA
Press COMMAND+PERIOD
Type an em dash
Press the PLUS SIGN (+)

Incorrect

SHIFT+ +
SHIFT+ -
Press +.

A B C D E F G H I J K L M N O P Q R S T U V W X Y Z

You can add the symbol in parentheses after the special character name—for example, PLUS SIGN (+). Use discretion in adding symbols, however; it may not be necessary for commonly used symbols such as PERIOD (.).

Microsoft Natural Keyboard key names

For conceptual topics and descriptions of programmable keys on the Microsoft Natural Keyboard, use the name of the key followed by its icon in parentheses at the first reference. For subsequent references, use the icon alone if possible. Note, however, that inline graphics affect line spacing, so using the name only may be preferable. The key name appears with an initial cap only, not in all caps.

Correct

You can define the Application key ( ) to open any program you want. Then press to open that program.
Then press the Application key to open that program.

Incorrect

Press the WINDOWS LOGO key.

Names of keyboard "quick access" keys

Differentiate among access keys and shortcut keys and use each term accurately. The following table lists the various types of key combinations and their meanings. See the specific topics for more details.

Keyboard "Quick Access" Keys

Name	*Alternative name*	*Definition*	*Audience*
Accelerator key	Shortcut key	Now obsolete in all uses.	Do not use.
Access key	Underlined letter	Underlined letter on a menu name, command, and so on, for keyboard access to the item.	Technical only. Use *underlined letter* in end-user documents.
Hot key		Key or key combination that activates a TSR (memory-resident program).	Use only with TSR programs
Quick key		Not a term that's used or defined.	Do not use.
Shortcut key		Key that corresponds to a command name on a menu, such as CTRL+Z.	All.
Speed key		Not a term that's used or defined.	Do not use.

Key Sequences

Use commas followed by spaces, as in ALT, F, D, to indicate that the user should press and release ALT, and then F, and then D.

SEE ALSO **Key Combinations, Key Names**

keypad

Always use *numeric keypad* for the first reference. Do not use *keypad* alone unless the context has been established and there's no possibility of confusion with the keyboard. If in doubt, continue to use *numeric keypad.*

In general, avoid making distinctions between the keyboard and the numeric keypad. When the user can press two keys that look the same, be specific in directing the user to the proper key—for example, "press the MINUS SIGN on the numeric keypad, not the HYPHEN key on the keyboard." Each group must resolve any problems this approach may cause because of the way certain keyboards and keypads function.

keypress

Do not use; use *keystroke* instead.

keystroke

One word. Not *keypress.*

Keywords and Online Index Entries

Current Windows indexes look similar to a print index: They are a two-level index with indented subentries. Likewise, most of the same conceptual guidelines for print indexes apply to online indexes.

The *keyword* is the term that a user associates with a specific task or set of information. The user types a keyword in the Find or Index box to locate specific information in a Help file. A keyword can lead to a single Help topic or to many related topics.

When deciding what keywords to list, consider these categories:

- Words for a novice user of your product
- Words for an advanced user of your product
- Common synonyms for words in the topics
- Words that describe the topic generally
- Words that describe the topic specifically
- Words commonly used by major competitors

Look specifically at these elements of your document for potential keywords when you develop your index:

- Headings
- Terms and concepts important to the user
- Overviews
- Procedures
- Acronyms
- Definitions or new terms
- Commands, functions, methods, and properties

Order of entries

Sort HTMLHelp indexes in the same way as print indexes. You cannot manually sort the Windows 95 keywords, so the order follows the ASCII sort order. Special characters appear first, then numbers, and then alphabetic entries.

Style of indexed keywords

Follow most of the same general style guidelines as those used for printed indexes:

- Use gerunds (the *ing* form) rather than infinitives (the *to* form) or the present tense base form of verbs for task-oriented entries, unless they are unsuitable, as they may be for languages, systems, or localized versions. Consult with your team when making this decision.
- Avoid generic gerunds that users are unlikely to look for as keywords: *using, changing, creating, getting, making, doing,* and so on.
- Use plural for nouns unless it is inappropriate. This applies to both single keywords *(bookmarks,* not *bookmark)* and keyword phrases *(copying files,* not *copying a file).*
- Avoid articles *(a, an, the)* and prepositions at the beginning of a keyword.
- Keep keywords as short as practicable for clarity.
- Use synonyms liberally, especially terms used in competitors' products or terms some users are likely to know: *option* and *radio button,* for example.
- For acronyms, list both the spelled-out phrase followed by the acronym in parentheses and the acronym followed by the spelled-out version: *terminate-and-stay-resident (TSR)* and *TSR (terminate-and-stay-resident).*
- Use all lowercase for all keywords and index terms unless the term is a proper noun or case-sensitive and thus capitalized in the Help topic.

Correct

clearing tab stops
clip art
Close command
modems
    dialing a connection manually
    setting up
    troubleshooting

Standardizing keywords across products

Some Help topics will be shared among products. Single-sourcing—using the same file for both the Help topic about copying and a book chapter about copying, for example—may extend across products instead of just occurring within a single product. This will require that some keywords also be standardized across products.

Topics, particularly those for technical support services and accessibility, should have standard keywords. For example, the technical support topic must include these keywords:

assistance, customer
customer assistance
help
Microsoft Technical Support
phone support
product support
support services
technical support
telephone support
troubleshooting

Merging keywords from multiple files

Windows 95 and later can present a single index for multiple Help files. The keywords from the separate Help files are merged as if the main contents (.cnt) file specifies each Help file. If such a Help system contains an optional component that the user does not install, those keywords will not show up in the index but will be added to the index if the user installs the component later.

A merged set of keywords can be very helpful for users. However, Help writers will need to make sure that the keywords fit together appropriately. For example, if the main Help file uses the phrase *quitting programs,* then all Help files in the project should use this phrase rather than just *quitting.* Otherwise, when the keywords from multiple files are merged, the user will see two entries, "quitting" and "quitting programs."

Cross-references

Avoid cross-references in online indexes. They are more difficult to handle in keyword lists than in print indexes.

Because each keyword must be linked to at least one Help topic, a cross-reference keyword has to jump somewhere, perhaps to an overview, or "main," topic. It is often difficult to determine which topic that should be.

Also, cross-references *(See* and *See also)* are limited to normal keywords that jump directly to the topic containing the K (keyword) footnote with that keyword. The cross-reference does not jump to another location in the index.

Instead of a cross-reference, duplicate all the subentries under both of the main keywords. For example, list all topics for "insertion point" under "cursor (insertion point)" as well.

If you must include a cross-reference to other topics, you will want to force it to the top of the list of subentries. Talk to your indexer about how to do this.

SEE ALSO **Indexing**

kHz

Abbreviation for *kilohertz.* Use the abbreviation only as a measurement with numerals; do not use in straight text without a numeral. Spell out *kilohertz* at first mention. See **kilohertz**.

kilobit

Always spell out. Do not use the abbreviation *Kbit.*

kilobits per second

Spell out at first mention; then use the abbreviation *Kbps.*

kilobyte

One kilobyte is equal to 1,024 bytes.

- Abbreviate as *KB,* not *K, K byte,* or *Kbyte.* At first mention, spell out and use the abbreviation in parentheses if your audience may not be familiar with the abbreviation.
- Separate the numeral from the abbreviation with a space or a hyphen, depending on usage.

  Correct
  800-KB disk drive
- When used as a noun in measurements, add *of* to form a prepositional phrase.

  Correct
  The Help files require 175 KB of disk space.

SEE ALSO **Measurements**

kilobytes per second

Spell out at first mention; then use the abbreviation *KBps.*

kilohertz

A kilohertz is a unit of frequency equal to 1,000 cycles per second.

- Abbreviate as *kHz.* At first mention, spell out and use the abbreviation in parentheses.
- Leave a space between the numeral and *kHz* except when the measurement is used as an adjective preceding a noun. In that case, use a hyphen.

  Correct
  The processor accesses memory at 500 kilohertz (kHz).
  a 900-kHz processor

SEE ALSO **Measurements**

kludge, kludgy

Slang. Do not use to refer to a Band-Aid fix or poorly designed program or system.

knowledge base, Knowledge Base

Use all lowercase for generic references to the "expert system" database type. Use title caps when referring to the Microsoft Knowledge Base. It's not necessary to precede *Knowledge Base* with the company name.

label, labeled, labeling

Do not double the final *l.*

landscape orientation

Printing orientation that prints horizontally across the wide side of the paper.

Landscape orientation

Compare **portrait orientation**.

laptop

One word. However, use *portable computer* in most instances because it does not refer to a specific size. See **portable computer**.

later

Use instead of *below* in cross-references—for example, "later in this section."

Use instead of *higher* for product version numbers—for example, "Windows version 3.0 or later."

SEE ALSO **Cross-References, higher**

launch

Do not use to mean *start,* as in "launch a program" or "launch a form." Use *start* instead.

SEE ALSO **start**

lay out (v), laid out (adj), layout (n)

Derivatives of *lay out* are commonly used in reference to formatting. Use the correct spelling and part of speech according to your meaning.

Correct

You can lay out complex information in a table.
Add formatting to your table after it is laid out.
A table layout clarifies complex information.

leave

Do not use to refer to quitting a program; use *quit* instead.

left

Not *left-hand.* Use *upper left* or *lower left, leftmost,* and so on. Include a hyphen if modifying a noun, as in "upper-left corner."

left align (v), left-aligned (adj)

Use to refer to text that's aligned on the left margin. Hyphenate the adjective in all positions in the sentence. Do not use *left-justified.*

left mouse button

In general, use just *mouse button;* use *left mouse button* only in discussions of multiple buttons or in teaching beginning skills.

SEE ALSO **mouse**

left-hand

Do not use; use just *left* instead.

SEE ALSO **left**

left-justified

Do not use; use *left-aligned* instead.

leftmost (adj)

One word. Use to refer to something at the farthest left side. Use instead of *farthest left, far-left,* or similar terms.

legacy (n)

Avoid using as an adjective, as in *a legacy system;* it's jargon. Instead, use *previous, former, earlier,* or a similar term. Describe the earlier systems if necessary, especially if discussing compatibility issues.

legal

Use only to refer to matters of law. Do not use to mean *valid,* as in "a valid action."

A B C D E F G H I J K L M N O P Q R S T U V W X Y Z

less vs. fewer vs. under

Use *less* to refer to a mass amount, value, or degree. Use *fewer* to refer to a countable number of items. Do not use *under* to refer to a quantity or number.

Correct

The new building has less floor space and contains fewer offices.
Fewer than 75 members were present.
Less than a quorum attended.

Incorrect

Less than 75 members were present.
The new building has less offices.
Under 75 members attended.
The new building has under 10 floors.

let, lets

Avoid in the sense of software permitting a user to do something. Use *you can* instead.

Correct

With Microsoft Project, you can present information in many ways.

Incorrect

Microsoft Project lets you present information in many ways.

SEE ALSO **can vs. may**

leverage (n)

Do not use as a verb; it's marketing jargon. Instead, use *take advantage of, capitalize on, use,* and so on.

like

Acceptable as a synonym for *such as* or *similar to.* Do not use as a conjunction; use *as* instead.

Correct

In a workgroup, you can work with files residing on another computer as you would on your own.
Moving a dialog box is like moving a window.

Incorrect

In a workgroup, you can work with files residing on another computer like you would on your own.

like (suffix)

In general, do not hyphenate words ending with *like,* such as *rodlike* and *maillike,* unless the root word ends in double *l*s (for example, *bell-like*) or has three or more syllables (for example, *computer-like*).

line

Do not use to refer to a series of related Microsoft products; use *family* instead.

Line Breaks

> **NOTE** This topic pertains primarily to printed documentation. Although you can set line breaks with the <BR> tag in HTML, most HTML documents use the default ragged right edge with no hyphenation. And because Help windows based on WinHelp can be sized, line breaks cannot be specified.

Although the right text edge in printed Microsoft documents is not aligned, try to avoid very short lines that leave large amounts of white space at the end of a line. An extremely ragged right edge can distract the reader. If necessary, a copy editor and desktop publisher can break lines manually during the final stages of production. For general rules about hyphenation and word division, see **Hyphens, Hyphenation**, *American Heritage Dictionary,* and *The Chicago Manual of Style.*

The following list gives the basic rules for line breaks in Microsoft printed documentation:

- Do not break a word if it leaves a single letter at the end of the line.
- Do not break a word if it leaves fewer than three letters at the beginning of the next line.

  Correct

  ```
  Be sure there are enough let-
  ters at the end of a line. Do
  not leave fewer than three
  letters at the begin-
  ning of a line.
  ```

  Incorrect

  ```
  Be sure there are e-
  nough letters at the end
  of a line. Do not leave few-
  er than three letters at
  the beginning of a line.
  ```

- Do not end a page with the first part of a hyphenated word.
- Avoid leaving fewer than four characters on the last line of a paragraph, especially if a heading follows.

A B C D E F G H I J K L M N O P Q R S T U V W X Y Z

- Do not hyphenate *Microsoft* unless there is no alternative. The only acceptable hyphenation is *Micro-soft.*
- Do not hyphenate Microsoft product names.
- Avoid breaking URLs. If you must break them, do so at the end of a section of the address immediately before the next forward slash. Do not include a hyphen.

  Correct
  For more information, connect to http://www.microsoft.com/support
  /products/developer/visualc/content/faq/

  Incorrect
  For more information, connect to http://www.microsoft.com/support-
  /products/developer/visualc/content/faq/
  For more information, connect to http://www.microsoft.com/support/
  products/developer/visualc/content/faq/

- Try to keep headings on one line. If a two-line heading is unavoidable, break the lines so that the first line is longer. Do not break headings by hyphenating words, and avoid breaking a heading between the parts of a hyphenated word. It doesn't matter whether the line breaks before or after a conjunction, but avoid breaking between two words that are part of a verb phrase.

  Correct
  Bookmarks, Cross-References,
  and Captions

  Incorrect
  Bookmarks, Cross-
  References, and Captions

- Try not to break formulas, data to be entered without spaces, or program examples. If a break is unavoidable, break between elements.

  Correct
  In the cell, type **=Budget!**
  AH:C#+1

  Incorrect
  In the cell, type **=Budget!$A**
  $H:$C$#+1

- Try to avoid breaking function names and parameters. If hyphenating is necessary, break these names between the words that make up the function or parameter, not within a word itself.

  Correct
  WinBroadcast-
  Msg

  Incorrect
  WinBroad-
  castMsg

- Do not hyphenate a line of command syntax or code. If you must break a line, break it at a character space, and do not use a hyphen. Indent the run-over when breaking a line of syntax or code. Do not use the line-continuation character unless it's necessary for the code to compile.

  Correct

  void CSribView::OnLButtonDown(**UINT** *nFlags, Cpoint*
  point)

  Incorrect

  void CScribView::OnLButtonDown(**UINT** *nFlags, C-*
  point point)

line feed (adj, n)

Two words. Refers to the ASCII character that moves the cursor or printing mechanism to the next line, one space to the right of its current position. Do not confuse with the *newline character,* which is the same as the carriage return/line feed and moves the cursor to the beginning of the next line. Can be abbreviated *LF* after the first use, as in *CR/LF.*

linking and embedding

Do not use this phrase; instead, use *OLE Linking and Embedding,* which are two features of OLE documents. You can, however, use phrases such as "linking information" and "embedding documents."

list box

Two words. *List box* is a generic term for any type of dialog box option containing a list of items the user can select. In text and procedures, refer to a list box by its label and the word *list,* not *list box.* For the Macintosh, use "pop-up list" to refer to unnamed list boxes.

List box

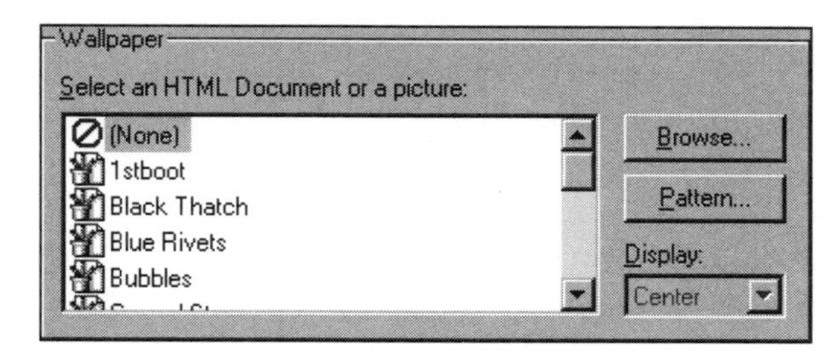

Correct

In the Wallpaper list, click Bubbles.
In the pop-up list, click Microsoft Excel. [Macintosh only]

Incorrect

In the Wallpaper list box, click Bubbles.

A
B
C
D
E
F
G
H
I
J
K
M
N
O
P
Q
R
S
T
U
V
W
X
Y
Z

Lists

Depending on how you want to present material, you can choose one of several types of lists: bulleted, numbered, unnumbered single-column, unnumbered multicolumn, or a "term list." A list can incorporate a nested comment, one nested list, or an untitled table.

A *table* is an arrangement of data with two or more columns in which the information in the first column relates horizontally to the information in the other column or columns. A list of similar entries that is arranged in multiple columns is not a table but a multicolumn list—for example, a list of commands. A table usually has column headings and may have a title. For more information, see **Tables**.

Punctuating lists

Introduce a list with a sentence or fragment ending with a colon. Begin each entry in a bulleted or numbered list with a capital letter.

Make entries in a list parallel. End each entry with a period if all entries are complete sentences, are a mixture of fragments and sentences, or complete the introductory sentence. An exception is when all entries are short imperative sentences (three words or fewer) or single words; these entries do not need a period. If all entries are fragments that do not complete the introduction, do not end them with periods.

Correct

If you use printer fonts:

- Choose a printer before creating a presentation.
- Install all the fonts and printers you'll use by selecting them in the Print Setup dialog box.

The database includes:

- Reports
- Forms
- Tables
- Modules

Bulleted lists

Use a bulleted list for an unordered series of concepts, items, or options rather than a sequence of events or steps. Capitalize the first word of each bulleted entry.

Correct

The database owner can:

- Create and delete a database.
- Add, delete, or modify a document.
- Add, delete, or modify any information in the database.

Numbered lists

Use a numbered list for procedures or other sequential lists. Introduce a procedure with an infinitive phrase or imperative, depending on your group's style, and avoid explanatory text after the phrase. Capitalize the first word of each entry.

Correct

To log on to a database

1. On the File menu, click Open Database.
2. In the User Name box, type your name.
3. In the Password box, type your password.
4. Click OK.

Correct

The basic steps for adding scrolling to your application are as follows:

1. Define a size for your documents.
2. Derive your view class from CScrollView instead of Cview.
3. Pass the documents' size to the SetScrollSizes member function of CScrollView whenever the size changes.
4. Convert between logical coordinates and device coordinates if you are passing points between GDI and non-GDI functions.

Unnumbered single-column and multicolumn lists

Use an unnumbered list to group similar items—for example, a list of keywords. Use a single column for six or fewer items and balanced multiple columns for seven or more. You need not capitalize entries. If the list is alphabetical, alphabetize down the columns, not across rows, if possible.

Because there are no page breaks in online files, long multicolumn lists can be difficult to read. In this case, you can alphabetize from left to right (for shorter lists) or sort in labeled alphabetical sections. Alphabetical sections make navigating in long lists of items such as functions easier.

Correct

and
or
nor
for

add	procedure
checkpoint	rule
errorexit	sum
nonclustered	triggerover

A-C

Abs	CDbl
Asc	Choose
Atn	Chr, Chr$
Avg	CIntCLng
Ccur	CodeDB

D-E

Date, Date$	ErrError, Error$
DateAdd	Erl
Environ, Environ$	EOF

Term lists

Use term lists for a series of terms, parameters, or similar items that are followed by a brief explanation or definition. In the HTML style sheets, the term appears on its own line, with the definition indented under it. The terms may use additional character formatting if appropriate. In other designs, the term is followed by a period, with the definition immediately following it.

NOTE These lists are also called "term-def" lists.

Correct

Computer name. The name by which the local area network identifies a server or workstation. Each computer name on the network must be unique.
Computer name
The name by which the local area network identifies a server or workstation. Each computer name on the network must be unique.

Incorrect

Computer name—The name by which the local area network identifies a server or workstation. Each computer name on the network must be unique.

SEE ALSO **Parallelism**

load

Do not use to mean *set up,* as in "setting up Word on a network file server." *Load* and *unload* (or *remove)* are used to describe adding (or removing) add-in programs.

SEE ALSO **download**

localization vs. translation

"Localization" refers to the process of making a program appropriate for the countries or part of the world in which it will be used. It includes changing the software itself as necessary. For example, sorting tables may have to follow a different character order in different languages.

"Translation" refers to the process of translating words or documents only into another language. Translation is just one part of localization.

Do not refer to a *localized version* in documentation. If necessary, it's acceptable to say something such as "If you are outside the United States and have a question about a product," followed by information about the list of subsidiary offices.

SEE ALSO **International Considerations**

lock

In general, do not use to mean *protect,* as in "protect a document from changes." Do not confuse with *write-protect,* which is what users do to disks to protect them from being overwritten. Some programs (for example, Microsoft Excel and Word) use *locked* to indicate portions of a document that can't be changed.

lock up

Avoid; instead, use *fail* for disks, *stop responding* for programs, and *shut down* for servers. It is acceptable to use *fail* to describe a disk or media failure, but *stop responding* should be used when discussing programs to avoid making it sound as if the software has failed.

log on to, log off from, logon (adj)

Use *log on to* to refer to connecting to a network and *log off from* (or simply *log off*) to refer to disconnecting from a network. Do not use *log in, login, log onto, log off of, logout, sign off,* or *sign on.* An exception is when other terms are dictated by the interface.

Use *logon* only as an adjective, as in "logon password," not as a noun.

Correct

You are prompted for your password while logging on.
Reconnect when you log on to the network.
Some networks support this logon feature.
Remember to log off from the network.

Incorrect

You are prompted for your password during logon.
Log in before you start Windows.
Remember to log off of the network.

Logical Operators

Also called *Boolean operators.* The most commonly used are **AND, OR, XOR** (exclusive **OR**), and **NOT.** Use bold all uppercase for logical operators. Do not use them as verbs, and do not use their symbols in text.

A
B
C
D
E
F
G
H
I
J
K
L
M
N
O
P
Q
R
S
T
U
V
W
X
Y
Z

Correct

Using **AND** to find *x* and *y* will produce **TRUE** only if both are true.

Incorrect

ANDing *x* and *y* produces **TRUE** only if both are true.
Using & to find *x* and *y* will produce **TRUE** only if both are true.

look at

Avoid; use *view* instead, as in "To view the list of Help topics, click Help."

look up

Acceptable to use instead of *see* in cross-references to online index entries from printed documentation. If you are using common source files for both printed and online documentation, however, use *see.*

Correct

For more information, look up "Dial-Up Networking" in the Help index.

lo-res

Do not use; use *low-resolution* instead.

lower

Do not use to indicate product version numbers. Use *earlier* instead, as in "Word version 3.0 or earlier." See **earlier**.

lower left (n), lower right (n)

Use instead of *bottom left* and *bottom right.* Hyphenate as adjectives: *lower-left* and *lower-right.*

lowercase (adj, n)

One word. Do not use *lowercased.* Avoid using as a verb.

When *lowercase* and *uppercase* are used together, do not use a suspended hyphen.

Correct

You can quickly change the capitalization of all uppercase and lowercase letters.
Change all the uppercase letters to lowercase.

Incorrect

You can quickly change the capitalization of all upper- and lowercase letters.
Lowercase all the capital letters.

low-level (adj)

Note hyphen. Use the term carefully; a *low-level language,* for example, means a language that is very close to machine language, such as assembly language.

SEE ALSO **high-level**

low-resolution

Note hyphen. Do not abbreviate as *lo-res.*

M

Do not use as an abbreviation for *megabyte.* See **MB**.

MAC

Acronym for *media access control.* However, do not use because of possible confusion with the Macintosh computer.

machine

Avoid; use *computer* instead.

machine language

Okay to use. Refers to the binary code that is the result of a program.

Macintosh

Do not shorten to *Mac.* According to the Apple publications style guide, *Macintosh* is most correctly used as an adjective, as in *Macintosh computer.* If you use it as a noun, always precede it with an article or possessive pronoun such as *the Macintosh* or *your Macintosh.*

Network language for the Macintosh differs from that of the personal computer. Use the terms *zone, file server,* and *shared disk* to refer to what users select to get information shared on a network. Use colons with no spaces to separate zones, file servers, shared disks, folders, and file names. File names have no extension.

Correct

Macintosh HD:My Documents:Sales
CORP-16:TOMCAT:EX130D Mac Temp:Workbook1

See other Macintosh topics listed in the index.

Macro Assembler

A programming language. Spell out at first occurrence; after that it can be abbreviated as *MASM.*

main document

Use to refer to the document that contains the unchanging material in a merged document, such as a form letter. Do not use *core document* or other terms.

makefile

Technical term. One word, lowercase.

management information systems

Abbreviate as *MIS,* but in general use *IS* for *information systems* instead, unless you know a specific reference should be to MIS.

manipulate

Avoid in end-user documentation. It's acceptable in technical documentation in reference to entities such as objects.

manual

In general, avoid *manual* as a synonym for *book, guide*, or other specific term referring to documentation. It sounds old-fashioned and is not user-friendly. Use the title of the book itself if possible.

SEE ALSO **Titles of Books**

Marginal Cross-References

Marginal cross-references, one type of marginal note that appears in print, can point out lessons or chapters in books or Help topics that give additional help or ideas.

You can follow standard cross-reference style, using a complete sentence and ending it with a period, or you can use a graphic (such as the Help button) with a heading and the cross-reference. Try to be consistent within product groups.

Correct

For more ideas, see
"Writing and Correspondence"
in *Getting Started.*

See *Getting Started*

The following are some basic guidelines for marginal cross-references:

- Avoid cluttering a page with too many marginal cross-references; some teams limit notations to three per page.
- Try to limit marginal cross-references to about three or four lines. They expand when localized.

- Break lines so that they are about the same length.
- Follow the design for a specific project to determine whether to apply character formatting in marginal notations.

SEE ALSO **Cross-References, Marginal Notes, Notes and Tips**

Marginal Notes

In printed documentation, marginal notes, often labeled "Tips," generally accompany procedures to give hints, shortcuts, or background information that the user should know before proceeding. These notes should be easy to read and should help minimize long text or additional steps within the procedure. You can also use marginal notes next to tables and art.

In Help, the equivalent of these notes can be a jump to another step, a pop-up window offering additional information such as a definition, or a tip at the end of a topic.

Begin a marginal note at the first step of the procedure and end it before or at the last step. If possible, place it next to the step it refers to.

You can include a heading to show the subject of the marginal note as shown in the first example.

Correct

About file names
Some restrictions apply to file names.
See Saving and Naming Files.

Tip If the Formatting toolbar isn't displayed, click Toolbars on the View menu, and then click Formatting.

The following are some basic guidelines for marginal notes:

- Avoid cluttering a page with too many marginal notes; some teams limit notations to three per page.
- Try to limit marginal notations to about three or four lines. They expand when localized.
- Break lines so that they are about the same length.
- Follow the design for a specific project to determine whether or not to apply character formatting in marginal notes.

SEE ALSO **Marginal Cross-References, Notes and Tips**

marquee

Okay to use to refer to the scrolling text feature on Web pages. Do not use to refer to the feature that draws a dotted line around a selection on the screen; use *bounding outline* instead.

SEE ALSO **dotted rectangle**

master/slave

Do not use in end-user documents. In programming documentation, it may be acceptable to refer to arrangements in which one device controls another as a *master/slave arrangement,* or to the controlling device as the *master* and the controlled device as the *slave.* This is common language in some computer uses and other technical industries, as in referring to switching stations in communications. However, because it could be offensive, make sure it's common in the area you are documenting before using it.

Do not use as a synonym for *parent/child.* These terms do not mean the same thing. See **parent/child**.

mathematical

Not *mathematic.*

matrix (s), matrices (pl)

Note spelling of plural; this is the preferred spelling in *American Heritage Dictionary.*

maximize

Acceptable as a verb.

Maximize button

Do not use *Maximize box* or *Maximize icon.* Refers to the button with an open square (Windows 95 and later) that is located in the upper-right corner of a window that has not been maximized. The Maximize button performs the same function as the Maximize command on a window's shortcut menu.

Use the phrase Maximize button to refer to the button, not just Maximize. It is acceptable to use "maximize" as a verb, however.

Correct

Click the Maximize button.
To fill the screen, maximize the window.

Click 

Incorrect

Click Maximize.

A B C D E F G H I J K L M N O P Q R S T U V W X Y Z

MB

Abbreviation for *megabyte.* Use the abbreviation only as a measurement with numerals; do not use in straight text without a numeral. Hyphenate the measurement when used as an adjective. Spell out *megabyte* at first mention. See **megabyte**.

Correct

4 megabytes (MB) of RAM
40-MB hard disk

SEE ALSO **Measurements**

Mbit, Mb

Do not use as abbreviations for *megabit;* always spell out.

Mbyte

Do not use as an abbreviation for *megabyte.* See **MB**.

Measurements

Avoid using measurements unnecessarily, especially in examples. When you do use measurements, follow these conventions:

- Use numerals for all measurements, even if the number is under 10. This is true regardless of whether the measurement unit is spelled out or abbreviated. Measurements include distance, temperature, volume, size, weight, points, and picas, but generally not units of time. Bits and bytes are also considered units of measure.

  Correct

  5 inches
  0.5 inch
  8 bits
  12 points high

- For two or more quantities, repeat the unit of measure.

  Correct

  3.5-inch or 5.25-inch disk
  64 KB and 128 KB

  Incorrect

  3.5- or 5.25-inch disk
  64 and 128 KB

- When a measurement is used as an adjective, use a hyphen to connect the number to the measurement. Otherwise, do not use a hyphen.

  Correct
  12-point type
  3.5-inch disk
  8.5-by-11-inch paper
  24 KB of memory

- Use *number* × *number,* not *number by number,* when discussing screen resolutions. If possible, use the multiplication sign, not a lowercase or uppercase *x.*

  Correct
  640 × 480 VGA

SEE ALSO **gigabyte, kilobyte, megabyte, terabyte**

Abbreviations of measurements

Avoid using abbreviations of most measurements when possible. Exceptions are abbreviations for kilobytes (KB), megabytes (MB), and gigabytes (GB). If you have space constraints, in a table for instance, use the following abbreviations.

Term	*Abbreviation*	*Notes*
Baud		Do not abbreviate *baud.*
Bits per second	bps	
Centimeters	cm	
Days	Do not abbreviate.	
Degrees	°	Usually temperature only.
	deg	Angle only.
Dots per inch	dpi	
Feet	ft	
Gigabits	Gbit	Do not use abbreviation; always spell out.
Gigabytes	GB	
Gigahertz	GHz	
Grams	g	
Hertz	Hz	
Hours	hr	
Inches	in. (or " [inch sign] if necessary)	
Kilobits	Kb	Do not use abbreviation; always spell out.
Kilobits per second	Kbps	Acceptable to use abbreviation Kb in this instance.
Kilobytes	KB	
Kilobytes per second	KBps	
Kilograms	kg	
Kilohertz	kHz	
Kilometers	km	
Lines	li	

(continued)

A B C D E F G H I J K L M N O P Q R S T U V W X Y Z

Term	*Abbreviation*	*Notes*
Megabits	Mb	Do not use abbreviation; always spell out.
Megabits per second	Mbps	Acceptable to use abbreviation Mb in this instance.
Megabytes	MB	
Megabytes per second	MBps	
Megahertz	MHz	
Meters	m	
Microseconds		Do not abbreviate *microseconds*.
Miles	mi	
Millimeters	mm	
Milliseconds	msec [or ms]	
Months	mo	Do not use abbreviation; always spell out.
Nanoseconds	ns	Do not use abbreviation; always spell out.
Picas	pi	
Points	pt	
Points per inch	ppi	
Seconds	sec [*or* s]	
Weeks	wk	Do not use abbreviation; always spell out.
Years	yr	Do not use abbreviation; always spell out.

Abbreviations of units of measure are identical, whether singular or plural—for example, *1 in.* and *2 in.*

When units of measurement are not abbreviated, use the singular for quantities of one or less, except with zero, which takes the plural *(0 inches).*

Use a space between the number and unit for all abbreviations. Also, close up *35mm* when used in a photographic context, as in "35mm slides."

Abbreviations of measurements appear without periods. An exception is the abbreviation for inch, which always takes a period.

Correct

1 point 1 pt
10 points 10 pt
1 centimeter 1 cm
1 inch 1 in.

medium (s), media (pl)

Follow conservative practice and use *medium,* not *media,* as a singular subject. However, *media* is now gaining acceptance as a singular when treated as a collective noun referring to the communications industry or community. If usage is unclear, be conservative but be consistent. Make sure the verb agrees with the subject (that is, *the medium is* and *the media are*), unless you are clearly using *media* as a collective noun in the singular.

Correct

The media include online broadcasts as well as newspapers, magazines, radio, and television.
The media covers news of the computer industry.

Media also refers to the physical material such as disks or CD-ROMs used to store electronic information.

Correct

The medium now used for many large computer programs is the DVD-ROM.

Do not use *media* as a shortened form of *multimedia.*

meg

Do not use as an abbreviation for *megabyte.* See **MB**.

megabit

Always spell out. Do not use the abbreviation *Mb* or *Mbit.*

megabits per second

Spell out at first mention; then use the abbreviation *Mbps.*

megabyte

One megabyte is equal to 1,048,576 bytes, or 1,024 kilobytes.

- Abbreviate as *MB,* not *M, meg,* or *Mbyte.* At first mention, spell out and use the abbreviation in parentheses.
- Leave a space between the numeral and *MB* except when the measurement is used as an adjective preceding a noun. In that case, use a hyphen.

  Correct

  1.2-megabyte (MB) disk
  1.2 MB
  40-MB hard disk

- When used as a noun in measurements, add *of* to form a prepositional phrase.

  Correct

  You can run many programs with only 1 MB of memory.

SEE ALSO **Measurements**

A B C D E F G H I J K L M N O P Q R S T U V W X Y Z

megahertz

A megahertz is a unit of frequency equal to 1 million cycles per second.

- Abbreviate as *MHz*. At first mention, spell out and use the abbreviation in parentheses.
- Leave a space between the numeral and *MHz* except when the measurement is used as an adjective preceding a noun. In that case, use a hyphen.

  Correct
  The processor accesses memory at 50 megahertz (MHz).
  90-MHz processor

SEE ALSO **Measurements**

memory

To avoid confusing users, refer to a specific kind of memory rather than use the generic term *memory,* which usually refers to random access memory (RAM). That is, use the more precise terms *RAM, read-only memory (ROM), hard disk*, and so on, as appropriate. It is all right to use *memory* for RAM if you're sure your audience will understand or if you've established the connection. In lists of hardware requirements, however, use *RAM.*

Follow the standard guidelines for using acronyms and abbreviating measurements such as kilobytes (KB) with reference to memory.

Correct
800-KB disk drive
The Help files require 175 KB of disk space.
Many applications now need at least 4 MB of RAM.

In the noun forms referring to memory measurements, use *of* in a prepositional phrase, as in "64 KB of RAM."

memory models

Do not hyphenate when referring to various memory models: *tiny memory model, large memory model,* and so on. Do hyphenate when the term modifies *program: tiny-memory-model program* and *large-memory-model program.*

memory-resident (adj)

Note hyphen. Use *memory-resident program,* not *TSR,* in end-user material. *TSR,* which stands for *terminate-and-stay-resident,* is acceptable for technical audiences.

menu item

Do not use in end-user documentation; use *command* instead. In programmer documentation about creating elements of a user interface, *menu item* may be the best term to use.

Menus and Commands

Menus contain commands. Dialog boxes contain command buttons and options. Do not refer to a command as a *menu item* (except in programming documents about interfaces), *a choice,* or an *option.*

> **NOTE** You can customize the menu bar and toolbar by adding additional menus, by putting menus on toolbars, and by dragging toolbar buttons to a menu. Customization does not change the essential characteristics of these items, however. They should still be treated as menus, commands, and so on.

Use *click* when referring to selecting or choosing commands, options, and dialog box buttons in procedures.

If, however, your group is documenting both mouse and keyboard instructions, you can use the generic *choose* or *select.* In this case, the user *selects* or *opens* menus; *chooses* commands that are *on* the menu; *selects* dialog box options; and *chooses* command buttons in dialog boxes. For information about dialog box terminology, see **Dialog Boxes and Property Sheets**.

Correct

On the File menu, click Open.
From the File menu, choose Open.

The following illustration shows elements of menus. The callouts use capitalization that's correct for the item, not for usual Microsoft callout style.

Menu with commands

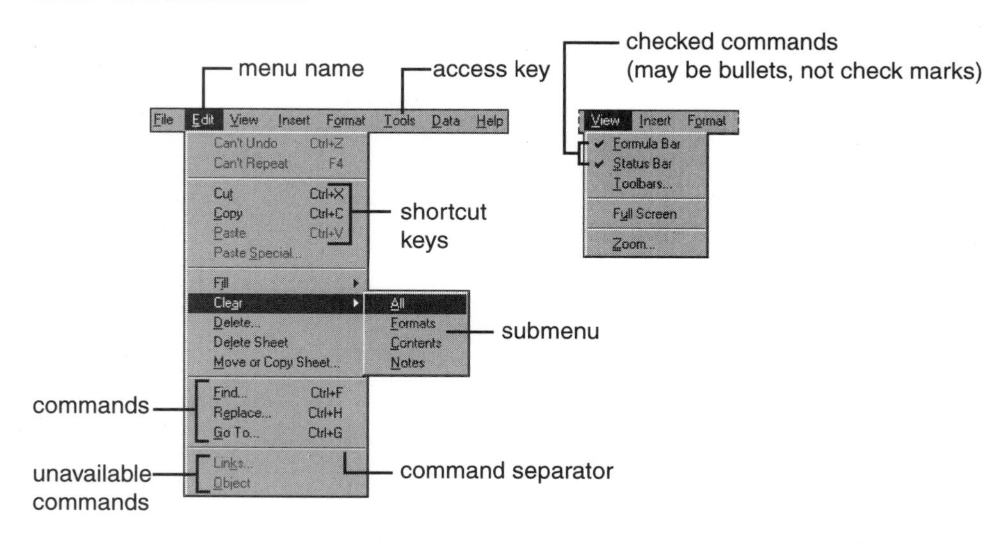

Menu terminology

- When referring to a specific menu, lowercase the word *menu,* as in “the Edit menu.”
- Avoid the words *cascading, pull-down, drop-down,* or *pop-up* to describe menus except in some programming documents. See **Types of Menus**, on page 175.
- Refer to unavailable commands and options as *unavailable,* not as *dimmed, disabled,* or *grayed.* In programming contexts, however, it's acceptable to refer to unavailable commands as *disabled.* Also, if you are describing the appearance of an unavailable command or option, you can use *dimmed,* but not *grayed* or *disabled.*

  Correct
  There are several unavailable commands on the Edit menu.
  If the Paste command is unavailable, first select the text you want to paste, and then choose Cut or Copy.
  The Paste command appears dimmed because it is unavailable.
  A disabled control is unavailable to the user.

  Incorrect
  There are several dimmed commands on the Edit menu.
  If the Paste command is disabled, first select the text you want to paste, and then choose Cut or Copy.
  The Paste command appears grayed because it is unavailable.

- Names of menus and menu commands are distinct elements on the screen. Do not combine the two names into one.

  Correct
  On the File menu, click Open.

  Incorrect
  Click File Open.

- In general, mention the name of the menu the first time you refer to a particular command. However, if the location of the command is clear from the immediate context (for example, a topic about the Edit menu), you may not need to mention the menu name.

  Correct
  In Control Panel, click the File menu, and then click Open.
  If the Paste command on the Edit menu is unavailable, first select the text you want to paste, and then choose Cut or Copy. You now should be able to choose Paste to insert the text in its new location.

- In Windows 95 and later, you open a submenu by *pointing* to the menu name.

  Correct
  Click the Start button, and then point to Documents.
  On the File menu, point to New, and then click Folder.

  Incorrect
  Click the Start button, and then choose Documents.
  On the File menu, choose New, and then choose Folder.

Types of menus

In end-user material, do not qualify the term *menu* with the adjective *drop-down, pull-down,* or *pop-up,* unless the way the menu works needs to be emphasized as a feature of the product. *Shortcut menu* is acceptable, however. Do not use any of these terms as verbs.

Correct

Open the File menu.
When you click the right mouse button, a shortcut menu appears.

Incorrect

Drop down the File menu.
When you click the right mouse button, a shortcut menu pops up.

In technical material, however, you might need to detail these specific kinds of menus to differentiate their programming constructions:

- Drop-down menu
- Pull-down menu (Macintosh documentation)
- Pop-up menu
- Shortcut menu
- Submenu

Style of menu names and commands

- Always surround menu names with the words *the* and *menu* both in text and procedures.

  Correct
  On the File menu, click Open.

  Incorrect
  On File, click Open.
  From File, click Open.

- In procedures, do not surround command names with the words *the* and *command.* In text, you can use "the ... command" for clarity.

  Correct
  On the File menu, click Open.

  Incorrect
  On the File menu, click the Open command.
  On the File menu, choose the Open command.

- Do not use the possessive form of menu and command names.

  Correct
  The Open command on the File menu opens the file.

  Incorrect
  The File menu's Open command opens the file.

A B C D E F G H I J K L M N O P Q R S T U V W X Y Z

- Follow the interface for capitalization, which usually will be title caps, and use bold formatting. Do not capitalize the identifier such as *menu* or *command.*

  Correct
  On the Options menu, click Keep Help on Top.

  Incorrect
  On the Options menu, click Keep Help On Top.

message (e-mail)

Use *message* to refer to the body of information in e-mail.

message box

In technical documentation, a secondary window that is displayed to inform a user about a particular condition. In end-user documentation, use *message.*

Messages

In end-user material, messages are online descriptions, instructions, or warnings that inform the user about the product or about conditions that may require special consideration. Refer to these simply as *messages,* not *alerts, error messages, message boxes,* or *prompts.* Include the term *error message* in indexes, however.

Error message is okay in technical documentation when describing types of messages.

When explaining a message, include the situation in which the message occurs, a paraphrase of the message (or, if necessary, the message itself), and what the user should do to continue. It is preferable to paraphrase a message because the wording might not be final in the software.

The text of a message should be short, clear, and to the point without being cryptic. The following table shows that a message can be in the form of a question or in passive voice to avoid casting blame on the user for an error. If possible, suggest a solution to a problem, as shown in the critical message example.

Interactive messages

An interactive message appears in a message box or in the Office Assistant (in Office 97) and requires a response to close it, such as clicking OK. For example, warning messages require the user to confirm an action before it is carried out. There are three types of interactive messages:

Symbol	*Name*	*Description*	*Example*
	Information	Provides information about the results of a command. Offers the user no choice.	Setup completed successfully.
	Warning	Informs the user about a situation that may require a decision, such as replacing an existing version of a document with a new one.	Do you want to save changes to Document 1?

(continued)

Symbol	Name	Description	Example
	Critical	Informs the user about a situation that requires intervention or correction before work can continue, such as a network being unavailable. Also used as the Stop button in Microsoft Internet Explorer.	The computer or sharename could not be found. Make sure you typed it correctly and try again.

NOTE Do not use the question mark symbol ( ) to indicate a message that appears in the form of a question. Users may confuse it with the Help symbol.

Informative messages

An informative message generally appears in the status bar at the bottom of the screen. An informative message can also appear in the Office Assistant in Office programs. A message in a program might tell the user the location within a document, for example. A command message in the status bar tells the user what the selected command will do.

Use present tense for informative messages that explain what a command does. The following example describes the New command on the Word File menu.

Correct

Creates a new document or template.

SEE ALSO **Error Messages, message box, prompt**

metadata

One word. Acceptable to use to refer to data that describes other data. A database term.

metafile

One word. Acceptable to use to refer to a file that describes or contains other files.

MHz

Abbreviation for *megahertz.* Use the abbreviation only as a measurement with numerals; do not use in straight text without a numeral. Spell out *megahertz* at first mention. See **megahertz**.

mice

Avoid; see **mouse**.

micro (prefix)

In general, do not hyphenate words beginning with *micro,* such as *microprocessor* and *microsecond,* unless it's necessary to avoid confusion or if *micro* is followed by a proper noun. If in doubt, check *American Heritage Dictionary* or your project style sheet.

microprocessor

Use instead of *processor* to refer to the chip used in personal computers.

Microsoft

At the first mention of Microsoft products, use the full name, such as *Microsoft PowerPoint.* After the first mention, you can shorten it to just *PowerPoint.* Never use the acronym *MS* for *Microsoft* with a product name.

Do not use the possessive form of a product name or a product's feature name. For example, do not use "toolbar's buttons" or "PowerPoint's File menu"; instead, use "toolbar buttons" and "the PowerPoint File menu."

Two products always require the use of *Microsoft:* Microsoft Project and Microsoft QuickBasic.

minicomputer

Do not abbreviate to *mini.*

minimize

Acceptable as a verb.

Minimize button

Do not use *Minimize box* or *Minimize icon.* Refers to the button containing a short line (Windows 95 and later) located in the upper-right corner of a window that has not been minimized. The Minimize button performs the same function as the Minimize command on a window's shortcut menu.

Use the phrase *Minimize button* to refer to the button, not just *Minimize*. It is acceptable to use "minimize" as a verb, however.

Correct

Click the Minimize button.
To reduce a program to a button on the taskbar, minimize the window.

Click

Incorrect

Click Minimize.

Minus Sign

Use an en dash for a minus sign except for user input when the user must type a hyphen or minus sign. In this case, the correct sign should be clearly noted in the documentation.

SEE ALSO **Em Dash, En Dash**

MIP mapping

Spell as shown. *MIP* is an acronym for "multum in parvo," Latin meaning "much in little." For more information, see the *Microsoft Press Computer Dictionary.*

monitor (n)

The television-like hardware that includes the screen; use *screen* to refer to the graphic portion of a monitor and *display* to refer to a visual output device (such as a flat-panel display on a portable computer).

Correct

Turn on the monitor.
A number of icons appear on the screen.
The newest portable computers have active-matrix LCD displays.

monospace

One word. A monospace font is used primarily for examples of code, including program examples and, within text, variable names, function names, argument names, and so on.

SEE ALSO **Code Formatting Conventions**, **Document Conventions**

Mood of Verbs

The mood of verbs expresses the "attitude" of the speaker toward the subject. The *indicative mood* makes general assertions, the *imperative mood* makes requests or commands, and the *subjunctive mood* expresses hypothetical information.

- The *indicative mood* is used to express general information such as facts, assertions, or explanations. It is also used to ask questions. Use this mood in most documentation.

  Correct

  Style sheets are powerful tools for formatting complex documents.
  What are the common characteristics of all interactors, including both text windows and scroll bars? They all have a size and relative position.

- The *imperative mood,* used for commands and requests, should be limited to procedures and direct instructions. The subject "you" is generally understood. Use only the present tense with the imperative mood.

  Correct

  Type a file name, and then click OK.
  Insert the disk in drive A.

- The subjunctive mood is used to express wishes, hypotheses, and conditions contrary to fact. Avoid this mood. It is seldom used today and is usually unnecessary in technical documentation.

  Correct (indicative mood)

  It is important to complete this procedure before taking any other action.

Incorrect (subjunctive mood unnecessary)
It is important that this procedure be completed before any other action is taken.

- Do not shift between moods.

  Correct
  Select the text, and then click the Bold button.
  First type a file name, and then click OK.

  Incorrect
  Select the text, and then you can click the Bold button.
  The first step is to type a file name, and then click OK.

more than vs. over

Use *more than* to refer to quantifiable figures and amounts. Use *over* to refer to a spatial relationship or position or nonquantifiable amounts.

Correct

The Design Gallery contains more than 16 million colors.
After you compress your drive, your disk will have over 50 percent more free space.
If you want the Help topic to appear over the document you are working on, click the On Top button.

SEE ALSO **over**

mouse

These sections describe how to refer to the mouse itself, how to use verbs such as *click, point,* and *drag* that refer to mouse actions, and how to handle mouse procedures.

Referring to the mouse correctly

- Avoid using the plural *mice;* if you need to refer to more than one mouse, use *mouse devices.* Never use *mice* in a specific reference to the Microsoft Mouse.
- Do not use *mouse cursor.*
- When describing the various ways the mouse pointer can appear on the screen, it is best to use a graphic. If this is not possible, it is acceptable to use descriptive labels for mouse pointers—for example, use *pointer,* not *double-headed arrow.*

  Correct

  When the pointer becomes a ←||→, use the arrow keys to move the split line.

- Use *right mouse button,* not other terms such as *mouse button 2, secondary mouse button,* and so on. Regardless of accuracy, users understand this term and users who reprogram their buttons make the mental shift.
- The Microsoft mouse that includes a wheel and wheel button is the "IntelliMouse pointing device." IntelliMouse is a registered trademark. Note capitalization. Always use *IntelliMouse* as an adjective, but do not use "the IntelliMouse mouse."
- Use *wheel button* to refer to the third (middle) button on the IntelliMouse pointing device.

Using mouse verbs correctly

- In general, use *point to,* not *move the mouse pointer to.* The latter is acceptable only in teaching beginning skills.

  Correct
  Point to the window border.

- Use *click,* not *click on.*

  Correct
  Using the mouse, click the Minimize button.

- Use *click* with a file, command, or option name, as in "click OK," but use *in* to refer to clicking in a general area within a window or dialog box, not *click the window* or *click the Styles box.*

  Correct
  To see the Control menu, click the right mouse button anywhere in the window.
  Click in the window to make it active.

  Incorrect
  To see the Control menu, use the right mouse button to click the window.

- Always hyphenate *double-click* and *right-click.*

  Correct
  Double-click the Word icon.
  Right-click to see the shortcut menu.

- Use *press and hold the mouse button* only to teach beginning skills.
- Use *drag,* not *click and drag* or *press and drag,* except in entry-level products. Note that *drag* includes holding down a button while moving the mouse and then releasing the button.
- Use *drag,* not *drag and drop,* for the action of moving a document or folder. It's okay to use *drag-and-drop* as an adjective, as in "moving the folder is a drag-and-drop operation."

  Correct
  Drag the folder to the desktop.

- Use *rotate,* not *roll,* to refer to rolling the IntelliMouse wheel.

  Correct
  Rotate the IntelliMouse wheel forward to scroll up in the document.

- In general, use *mouse button;* use *left mouse button* only to teach beginning skills.
- Use right-click to mean "click with the right mouse button" after you have clarified the meaning.

  Correct
  You can click with the right mouse button (called *right-click*) to see a shortcut menu.

- When more than one mouse button is used within a procedure, identify only the least commonly used button.

  Correct
  With the right mouse button, double-click the icon.

A B C D E F G H I J K L M N O P Q R S T U V W X Y Z

Handling mouse procedures

- Be consistent in the way you list mouse procedures; for example, always list the mouse method before listing the keyboard method.
- Do not combine keyboard and mouse actions such as key sequences.

  Correct
  Hold down SHIFT and click the right mouse button.

  Incorrect
  SHIFT+click the right mouse button.

movable

Not *moveable.*

movement keys

Do not use; use *arrow keys* instead. See **Key Names**.

MPEG

Abbreviation for Moving Picture Experts Group (sometimes seen as Motion Pictures Experts Group), a standard for storing motion pictures in compressed form. Files in MPEG format are used on CD-ROMS, video CDs, and DVDs. The extension for MPEG files is .mpg.

For more information, see the *Microsoft Press Computer Dictionary.*

MS

Do not use as an abbreviation for Microsoft. See **Microsoft**.

MS-DOS

Do not use *DOS.* Do not use *MS-DOS* as an adjective before anything that is not a component or aspect of the MS-DOS operating system; use *MS-DOS-based* instead.

Correct

MS-DOS-based program
MS-DOS-based computer
MS-DOS command

Incorrect

DOS program
MS-DOS program
MS-DOS computer

SEE ALSO **DOS**

MS-DOS file attributes

Use lowercase for MS-DOS file attribute names, as in "the read-only, archive, system, and hidden file attributes."

MS-DOS prompt

Avoid; use *command prompt* instead.

MS-DOS-based program

Use instead of *non-Windows program* when discussing software that runs on the MS-DOS operating system. *Character-based application* is acceptable for generic references to programs that do not run in Windows or other graphical environments, if the audience is familiar with the term.

Note that the term is spelled with two hyphens, not a hyphen and an en dash.

multi (prefix)

In general, do not hyphenate words beginning with *multi* unless it's necessary to avoid confusion or if *multi* is followed by a proper noun. Avoid inventing new words by combining them with *multi.* If in doubt, check *American Heritage Dictionary* or your project style sheet. If the word does not appear, use *multiple* before the word instead.

Correct

multicast
multichannel
multicolumn
multilevel
multiline
multilingual
multimedia
multiprocessor
multipurpose
multitasking
multiuser

multiple selection

Not *disjoint selection* or *noncontiguous selection.* It is okay to use *nonadjacent selection* to differentiate between any multiple selection and one in which you want to emphasize a separation.

Avoid using the verbal phrase "multiply select." Instead, use a phrase such as "you can select multiple items."

multiplication sign (×)

In general, use the multiplication sign (×), not an *x,* to indicate the mathematical procedure. Use an asterisk (*) if required to match the software. Also use × to mean "by" in referring to screen resolution.

Multiplication sign

×

SEE ALSO **Measurements**

multitasking (adj, n)

Do not use verb forms of this word; it is jargon.

Correct

Windows 95 supports multitasking.

Incorrect

You can multitask with Windows 95.

multithreaded (adj)

Not *multithread.*

My Computer

A Windows 95 and later icon that represents a user's private, local system. To refer to the icon in printed documents, use just My Computer if the icon is shown; otherwise, use "the My Computer icon."

My Computer

Correct

My Computer is visible at the upper left of your Windows desktop.
Double-click the My Computer icon.

n

Conventionally, a lowercase italic *n* refers to a generic use of a number. You can use *n* when the value of a number is arbitrary or immaterial—for example, "move the insertion point *n* spaces to the right." Reserve italic *x* for representing an unknown in mathematical equations (a variable) and other such placeholders.

SEE ALSO **x**

nanosecond

One-billionth of a second. Always spell out. Do not use the abbreviation *NS* or *ns.*

native language

Avoid when referring to a computer system's machine language; this term could be a misleading anthropomorphism. Instead, use *machine language* or *host language.*

navigate

Avoid the verb *navigate* to refer to moving from site to site, page to page within a site, or link to link on the Internet (or on the desktop or in other applications, as well). Instead, use *explore* to mean looking for sites or pages generally, *move to* or *move through* to refer to sequentially moving from one link or site to another, or a similar neutral term describing the action.

Navigate, navigating, and *navigation,* long-established terms used to refer to moving through a screen or program, are acceptable to refer to controls or buttons on the interface (for example, "navigation buttons") or as Help or topic references.

need

Often confused with *want.* Be sure to use the term that is appropriate to the situation. *Need* connotes a requirement or obligation; *want* indicates that the user has a choice of actions.

Correct

If you want to use a laser printer, you need a laser printer driver.

A B C D E F G H I J K L M N O P Q R S T U V W X Y Z

Negative Construction

State information positively whenever possible. Tell the user what to do and the reason, not what *not* to do. Using *not* can make a sentence harder to read and understand.

Positive

Save your work, and then turn off your computer.
It is possible to lose all your work, so back it up to be safe.

Negative

Do not turn off your computer without saving your work.
It is not impossible to lose all your work, so back it up to be safe.

However, you may need to use a negative construction for emphasis, as in a caution or warning, for example.

Acceptable use of the negative

Caution

Do not turn off your computer without saving your work.

Net

Slangy abbreviation for "Internet," sometimes acceptable in marketing materials. Do not use in documentation.

network (n), networking (n)

Do not use *network* as a verb, and do not shorten to *net* (jargon). A computer is *on* a network (not *in),* and computers are linked or connected, not *networked.*

Use the noun *networking* only to refer to making personal and business connections.

network adapter

You can use this term or the alternative *network interface card* (NIC) to refer to the printed circuit board in client and server computers that allows them to exchange data. *Network adapter* is more general and preferable in end-user documents, but whatever your choice, use it consistently and precisely.

network administrator

Use only to specifically refer to the administrator of networks. In general, use *administrator* or *system administrator* unless you must specify a particular kind.

SEE ALSO **system administrator, sysop**

network connection

Not *local area network connection.*

A B C D E F G H I J K L M N O P Q R S T U V W X Y Z

network drive

Not *remote drive.*

newsgroup

An Internet discussion group focusing on a particular topic. Use instead of *group* in this sense. Reserve *group* for more generic uses.

new line (adj + n), newline (adj)

Use two words to refer to the result of starting a new line. Use the one-word adjective only to refer to the ASCII end-of-line code (CR/LF), which moves the insertion point to the beginning of a new line. Use *newline character* instead of *end-of-line mark* to refer to the ASCII end-of-line code.

Correct

Press SHIFT+ENTER to start a new line.
Use the newline character to move to the beginning of the next line.

non (prefix)

In general, do not hyphenate words beginning with *non,* such as *nonalphanumeric, nonnumeric,* and *nonzero,* unless it's necessary to avoid confusion or *non* is followed by a proper noun, as in *non-English.* If in doubt, check *American Heritage Dictionary* or your project style sheet.

noncontiguous selection

In end-user documentation, do not use to refer to a type of multiple selection in which the items (such as cells in a table or worksheet) do not touch; instead, use *multiple selection.* If you must specify the type of multiple selection, use *adjacent selection* or *nonadjacent selection.* In technical documentation, avoid unless the term is used in the product.

Correct

To select multiple adjacent cells, drag across the cells that you want to select.
To select nonadjacent cells, select a single cell, and then hold down the CTRL key while you click other cells that you want to select.

SEE ALSO **adjacent selection**

nonprintable, nonprinting

Use *nonprintable* to refer to an area of a page that cannot be printed on. Use *nonprinting* to refer to characters and other data that can't or won't be printed. Do not use *unprintable,* which means "not fit to be printed."

Correct

Some text extends into the nonprintable area of the page.
When you click Show/Hide, Word displays all nonprinting characters, including paragraph marks and space marks.

A B C D E F G H I J K L M N O P Q R S T U V W X Y Z

Nonsexist Language

Use neutral forms of words rather than masculine forms (*workforce,* not *manpower; sales representative,* not *salesman).* Rewrite to avoid using the generic masculine pronouns *he* and *his.* For more information, see **Bias-Free Communication**.

non-Windows application, non-Windows-based

Do not use; use the names of specific operating systems instead, such as *MS-DOS-based program, UNIX program,* and so on.

normal, normally

Implies "in a normal manner," which may not be possible for everyone. Do not use to mean "often," "usual," or "typical." Instead, use *usually, ordinarily, generally,* or a similar term.

Notes and Tips

Notes (including cautions, important notes, tips, and warnings, as well as general notes) call the user's attention to information of special importance or information that can't otherwise be suitably presented in the main text. Use notes sparingly so that they remain effective attention-getters.

Material cluttered with notes probably indicates a need to reorganize the information. In general, try to use only one note in a Help topic. For example, if you must have two notes, combine them into one with two paragraphs. In any case, do not use two or more separate notes without intervening text.

You can include lists within notes.

The type of note (distinguished by the heading of the note or its bitmap) depends on the type of information given, the purpose of the information, and its relative urgency. The following sections explain the types of notes possible and their rank, from neutral to most critical.

Notes

A *note* with the heading "Note" indicates neutral or positive information that emphasizes or supplements important points of the main text. A note supplies information that may apply only in special cases—for example, memory limitations, equipment configurations, or details that apply to specific versions of a program.

Correct

Note If Windows prompts you for a network password at startup, your network is already set up and you can skip this section.

There is no symbol or bitmap to indicate a note.

Tips

A *tip* is a type of note that helps users apply the techniques and procedures described in the text to their specific needs. A tip suggests alternative methods that may not be obvious and helps users understand the benefits and capabilities of the product. A tip is not essential to the basic understanding of the text.

Correct

Tip You can also use this procedure to copy a file and give it a new name or location.

In printed documentation, the tip icon signals a tip, with or without the heading "Tip." In most Help topics no tip bitmap appears, just the word, as shown in the preceding example.

Important notes

An *important note* is a type of note that provides information essential to the completion of a task. Users can disregard information in a note and still complete a task, but they should not disregard an important note.

Correct

Important The device drivers installed automatically during Setup are required by your system. If you remove one of these drivers, your system may not work properly.

There is no symbol or bitmap for an important note.

Cautions

A *caution* is a type of note that advises users that failure to take or avoid a specified action could result in loss of data.

Correct

Caution To avoid damaging files, always shut down your computer before you turn it off.

In online messages, the Warning symbol indicates a caution. There is no symbol in print for a caution.

Warnings

A *warning* is a type of note that advises users that failure to take or avoid a specific action could result in physical harm to the user or the hardware. Use a warning, not a caution, when physical damage is possible.

Correct

Warning Do not let your Microsoft Mouse come in contact with water or other fluids. Excessive moisture can damage the internal circuitry of the mouse.

The warning symbol is used for both printed and online documentation.

A B C D E F G H I J K L M N O P Q R S T U V W X Y Z

Nouns

Do not turn verbs into nouns or create verbs from nouns. Try to use concrete nouns and active verbs to express the meaning of a sentence.

Correct

You can find one term in the document and replace it with another.
We can reuse this material in the next release.

Incorrect

You can do a find-and-replace on the document.
We can leverage this material for the next release.

SEE ALSO **Verbs**

NT

Do not use; always use *Windows NT.* This is a Microsoft trademark.

NUL, null, NULL, Null

Be sure to preserve the distinction between a *null* (ASCII *NUL) character* and a *zero character.* A null character displays nothing, even though it takes up space. It is represented by ASCII code 0. A zero character, on the other hand, refers to the digit 0 and is represented by ASCII code 48.

Use lowercase *null* to refer to a null value. Better yet, use *null value* to avoid confusion with the constant.

Use *NULL* or *Null* (depending on the language) only to refer to the constant.

null-terminated (adj), null-terminating

Use *null-terminated* as an adjective, as in "null-terminated string." Do not use *null-terminating,* as in "null-terminating character"; use *terminating null character* instead.

number sign (#)

Use *number sign,* not *pound sign,* to refer to the # symbol. It is acceptable, however, to use *pound key (#),* including the symbol in parentheses, when referring specifically to telephones or telephone numbers.

Always spell out *number;* do not use the # symbol (except as a key name)—for example, use *number 7,* not *#7.* When necessary to save space, as in tables, the abbreviation *No.* is acceptable.

Numbers

The sections in this topic discuss when to use numerals and when to spell out numbers, how to treat fractions and ordinal numbers, when to use commas in numbers, and how to treat ranges of numbers.

Numerals vs. words

The use of numerals versus words is primarily a matter of convention. Microsoft uses the following conventions:

- Use numerals for 10 and above. Spell out zero through nine if the number does not precede a unit of measure or is not used as input. For round numbers of 1 million or more, use a numeral plus the word, even if the prefix number is less than 10.

  Correct
  10 screen savers
  20 booklets
  1,000
  one thousand
  five databases
  zero probability
  7 million
  7,990,000

  Incorrect
  2 disks
  0 offset
  eighteen books
  twelve functions
  1 thousand
  7 million and 990 thousand

  Use numerals for all measurements, even if the number is under 10. This is true whether the measurement is spelled out, abbreviated, or replaced by a symbol. Measurements include distance, temperature, volume, size, weight, points, picas, and so on, but generally not days, weeks, or other units of time. Bits and bytes are also considered units of measure.

  Correct
  0 inches
  3 feet, 5 inches
  3.5-inch disk
  0.75 gram
  35mm camera
  8 bits
  1-byte error value
  two years

- Use numerals in dimensions. In most general text, spell out *by,* except for screen resolutions. For those, use the multiplication sign (×).

  Correct
  8.5-by-11-inch paper
  640 × 480

A B C D E F G H I J K L M N O P Q R S T U V W X Y Z

- Use numerals to indicate the time of day. Use a 12-hour clock for domestic documentation, including the minutes, followed by A.M. or P.M. To avoid confusion, use 12:00 noon or 12:00 midnight for those precise times.

  Correct
  10:00 P.M.
  12:01 A.M.

  Incorrect
  nine o'clock
  10 P.M.
  12:00 A.M.

- Spell out the name of the month in dates because the position of the month and day are reversed from the U.S. system in other systems.

  Correct
  June 12, 1996
  December 6, 1996

  Incorrect
  6/12/96
  12/6/96

- Maintain consistency among categories of information; that is, if one number in a category requires a numeral, use numerals for all numbers in that category. When two numbers that refer to separate categories must appear together, spell out one of them.

  Correct
  One booklet has 16 pages, one has 7 pages, and the third has only 5 pages.
  ten 12-page booklets

- Use numerals for coordinates in tables or worksheets and for numbered sections of documents.

  Correct
  row 3, column 4
  Volume 2
  Chapter 10
  Part 5
  step 1

- Represent numbers taken from examples or the interface exactly as they appear in the example or the interface.
- Use an en dash, not a hyphen, with negative numbers: –79.
- Avoid starting a sentence with a numeral. If necessary, add a modifier before a number. If starting a sentence with a number cannot be avoided, spell out the number.

Correct
Lotus 1-2-3 presents options in the menu.
Microsoft Excel has 144 functions.
Eleven screen savers are included.
The value 7 represents the average.

Incorrect
1-2-3 presents options in the menu.
144 functions are available in Microsoft Excel.
11 screen savers are included.
7 represents the average.

- Hyphenate compound numbers when they are spelled out.

  Correct
  Twenty-five fonts are included.
  the forty-first user

Fractions as words and decimals

Express fractions in words or as decimals whenever possible, whichever is most appropriate for the context.

- Hyphenate spelled-out fractions used as adjectives or nouns. Connect the numerator and denominator with a hyphen unless either already contains a hyphen.

  Correct
  one-third of the page
  three sixty-fourths
  two-thirds completed

- In tables, align decimals on the decimal point.
- Use an initial zero for decimal fractions less than one. When representing user input, however, do not include a zero if it is unnecessary for the user to type one.

  Correct
  0.5 inch
  type .5 inch

- When units of measure are not abbreviated, use the singular for quantities of one or less, except for zero, which takes the plural.

  Correct
  0.5 inch
  0 inches
  5 inches

- If an equation containing fractions occurs in text, you can use the Word Equation Editor to format it. Or, to insert a simple fraction, use a slash mark (/) between the numerator and the denominator.

  Correct
  1/2 + 1/2 = 1

Ordinal numbers

Ordinal numbers designate the place of an item in a sequence, such as *first, second,* and so on.

Cardinal numbers	*Ordinal numbers*
One, two	First, second
31, 32	Thirty-first, thirty-second
161	One hundred sixty-first

- Spell out ordinal numbers in text.

  Correct
  The line wraps at the eighty-first column.

  Incorrect
  The line wraps at the 81st column.

- Do not use ordinal numbers for dates.

  Correct
  The meeting is scheduled for April 1.

  Incorrect
  The meeting is scheduled for April 1st.

- Do not add *ly,* as in *firstly* and *secondly.*

Commas in numbers

In general, use commas in numbers that have four or more digits, regardless of how the numbers appear in the interface. When designating years and baud, however, use commas only when the number has five or more digits.

Do not use commas in page numbers, addresses, and decimals.

Correct

1,024 bytes
page 1091
1,093 pages
1.06377 units
10,000 B.C.
9600 baud
14,400 baud

Incorrect

page 1,091
2492 days
4,400 Park Avenue
10000 B.C.
9,600 baud

Ranges of numbers

Use *from* and *through* to describe inclusive ranges of numbers most accurately, except in a range of pages, where an en dash is preferred. Where space is a problem, as in tables and online material, use an en dash to separate ranges of numbers. You can use hyphens to indicate page ranges in an index if you need to conserve space.

Do not use *from* before a range indicated by an en dash. Avoid using *between* and *and* to describe an inclusive range of numbers because it can be ambiguous.

Correct

from 9 through 17
1985–1990
pages 112–120

Incorrect

between 9 and 17
from 1985–1990

SEE ALSO **less vs. fewer vs. under, more than vs. over**

numeric

Not *numerical.*

Also, use *numeric keypad,* not *keypad, numerical keypad,* or *numeric keyboard.*

SEE ALSO **keypad**

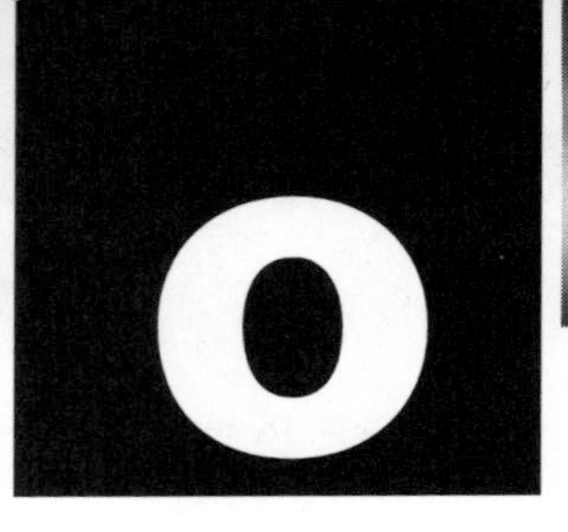

object (n)

Avoid using *object* as a synonym for "item" or "thing." Try to be as specific as possible when you refer to an object, because the term means different things in different programs.

For example, in C++ programming, an "object" is an instance of a "class" (a kind of user-defined type). It contains both routines and data and is treated as one entity. Similarly, in COM-based technologies, an object is a combination of code and data that implements one or more interfaces. In assembly language, however, it refers to the object module, which contains the data that's been translated into machine code.

SEE ALSO **COM, ActiveX, and OLE Terminology, embed**

object linking and embedding

Do not use as the spelled out meaning of *OLE.* It is acceptable to use generically and to use other grammatical forms of the verbs *(link, linked).*

SEE ALSO **COM, ActiveX, and OLE Terminology**

obsolete (adj)

Do not use as a verb. It's slang. Use a phrase such as "make obsolete" instead.

of

Do not use *of* after another preposition—for example, "off of" or "outside of." It's colloquial and unnecessary.

Correct

The taskbar is outside the main window area.
Save your work and then log off the network.

Incorrect

The taskbar is outside of the main window area.
Save your work and then log off of the network.

offline (adj, adv)

One word in all instances. Use in the sense of not being connected to or part of a system. It's jargon in the sense of "outside the current considerations."

okay, OK

Use *OK* only to match the interface; otherwise, use *okay.* When referring to the OK button, don't use *the* and *button.*

Correct

In the Save As dialog box, click OK.
It's okay to use more than eight characters to name a file in Windows 95.

Incorrect

In the Save As dialog box, click the OK button.

on

Use *on* with these elements:

- Menus ("the Open command is on the File menu")
- Taskbar, toolbar, ruler, and desktop ("click Start on the taskbar")
- Disks, in the sense of a program being on a disk ("the printer drivers on Disk 4")
- Interface ("on the interface")
- The screen itself (something appears "on the screen")
- Network ("the printer is on the network")
- Hardware platforms ("on the Macintosh")

Do not use *on* with user input actions.

Correct

Click the right mouse button.
Click the WordPad icon.
Click OK.
Press ENTER.

Incorrect

Click on the right mouse button.
Click on the WordPad icon.
Click on OK.
Press on the ENTER key.

SEE ALSO **in, into, on-screen, onto, on to, Procedures**

A B C D E F G H I J K L M N O P Q R S T U V W X Y Z

once

To avoid ambiguity, do not use as a synonym for *after.*

Correct

After you save the document, you can quit the program.

Incorrect

Once you save the document, you can quit the program.

online (adj, adv)

One word in all instances. Avoid the word if possible, however, because it now seems synonymous with Internet applications rather than more generically contrasting computerized material with other media such as print. Try to be specific or clarify the meaning of *online*.

Correct

Many Microsoft support services are available online through the World Wide Web.
Many products include online documentation on the CD-ROMs in the package.

online Help

Avoid the redundancy of using *online* except when necessary to describe the Help system; in general, use just *Help*.

Correct

You have easy access to hundreds of subjects in Help.

Incorrect

You have easy access to hundreds of subjects in online Help.

on-screen (adj, adv)

Hyphenate as both an adjective and adverb in all instances. However, instead of using it as an adverb, try to write around by using a phrase such as "on the screen."

Correct

Follow the on-screen instructions.
Follow the instructions that appear on the screen.
The instructions on the screen will not print.

on the fly

Okay to use in programming documentation to refer to a process that occurs without disrupting or suspending normal operations. For more information, see the *Microsoft Press Computer Dictionary.*

onto, on to

Use two words *(on to)* for the action of connecting to a network, as in "log on to the network."

Use one word *(onto)* to indicate moving something to a position on top of something else, as in "drag the icon onto the desktop."

on/off switch

Okay to use. Do not use *on/off button* except when referring to a remote control device.

open (adj, v)

Users open windows, files, documents, and folders. They can click or choose an item to open it. Do not use *open* to describe choosing a command, a menu, an icon, an option, or other similar element that doesn't produce a working file in a window. Describe the item as *open,* not *opened,* as in "an open file" and "the open document."

Correct

To open the document in Outline view, click View, and then click Outline.
You double-click the Works icon to open Works.
You can view your document in the open window.

Incorrect

Open the View menu, and then open the Outline command.
Open the Works icon.
You can view your document in the opened window.

SEE ALSO **Menus and Commands, Procedures**

opcode

Avoid using as a shortened form of *operation code*. It's jargon.

operating environment, operating system

Conventionally, an *operating environment* (or just *environment*) includes both hardware and operating system software, while an *operating system* comprises the software only. (A *graphical environment* refers to the graphical interface of an operating system.) In practical use, however, *environment* often refers only to the operating system, as in "FoxPro runs in the UNIX environment."

Although the term *platform* refers specifically to hardware architecture, it is sometimes used interchangeably with *environment.* Avoid this jargon in documentation.

Various prepositions are acceptable to use with *operating system:* Programs can run *with, on,* or *under* an operating system, whichever seems more appropriate. However, do not use *run against* an operating system in any kind of documentation.

Correct

Word 6.0 runs with both Windows and Macintosh operating systems.
Microsoft Exchange Server runs on Windows NT.

Incorrect

A number of programs run against Windows 95.

SEE ALSO **platform**

option (n), option button

In both end-user and technical material, use *option* to refer to the choices in dialog boxes. Capitalize the name of an option, following the interface style, but do not capitalize the word *option* itself.

It is usually unnecessary to refer to the button itself, however, only to the option it controls.

Correct

In the Sort Text dialog box, click No Header Row.

Incorrect

In the Sort Text dialog box, click the No Header Row option button.

Option button refers to the round button that indicates choices in a dialog box. Avoid *radio button* except to mention as a synonym and in indexes.

Option buttons

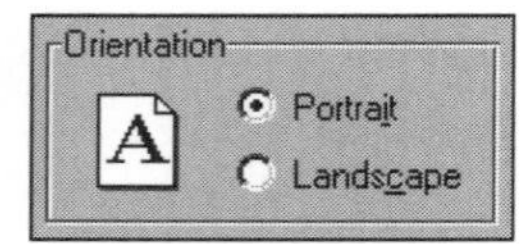

In technical material, you can use *option* instead of *switch* to refer to a command argument or compiler option such as /b or /Za. Refer to your project style sheet.

SEE ALSO **control, Dialog Boxes and Property Sheets, switch**

Ordinal Numbers

Ordinal numbers designate the place of an item in a sequence: *first, second,* and so on. Spell out in text, even when more than nine; that is, do not use *1st, 2nd, 12th,* and so on. Do not add *ly,* as in *firstly* and *secondly.*

SEE ALSO **Numbers**

outdent

Do not use unless it is explicitly a part of a product's user interface (such as Microsoft Project or Microsoft Access); instead, use "extend text into the margin."

SEE ALSO **indent**

output (n, adj)

Do not use as a verb; instead, use a term specific to the kind of output referred to, such as *write to, display on,* or *print to,* not *output to.*

Correct

The output provided the information needed.
A printer is a standard output device.
You can print a document to a file instead of to a specific printer.

Incorrect

You can output a document to a file instead of to a specific printer.

outside

Use instead of the colloquial *outside of.*

over

To avoid ambiguity, use *over* to refer to a position or location above something. For quantities, use *more than.* Do not use to refer to version numbers; instead, use *later.*

Correct

The [Installable ISAMs] heading appears over the list of paths to ISAM drivers.
The installed base is more than 2 million.
You need Windows 3.1 or later.

Incorrect

The installed base is over 2 million.
You need Windows 3.1 or over.

overtype

Acceptable if this term is used in the product's user interface. *Type over* (two words) is also acceptable, as long as it cannot be confused with *redo.*

Capitalize when referring to *Overtype mode.*

A B C D E F G H I J K L M N O P Q R S T U V W X Y Z

overwrite

Avoid, except to refer to the Overwrite mode or in the sense of recording new data on top of existing data. In procedural information, use *overtype* or *type over* if you're referring to text or *replace* if you're referring to files.

Correct

If you press the INSERT key, you can type over existing text.
Replace the previous document with the corrected one.

Incorrect

If you press the INSERT key, you can overwrite existing text.
Overwrite the previous document with the corrected one.

page

Refers to one of a collection of Web documents that make up a Web site. Use *page* to refer to the page the user is on, that is, the particular document, or to a specific page such as the home page or start page.

Page Breaks

Do not break pages in print documents until all art (or art spaces) and textual changes have been added to the manuscript. This step usually occurs just before the manuscript goes to final production.

The main goal is to keep related material on one page. If this isn't feasible, try to break pages so the user knows that relevant material continues on the next page. Recto (right) page breaks must be handled more carefully than those on a verso (left) page of a spread. Avoid leaving a recto page so short that it looks like the end of a chapter.

- Leave at least two lines of a paragraph at the bottom or top of a page. Do not break a word over a page.
- Avoid separating notes, tips, important notes, cautions, and warnings from the material they concern.
- Keep material introducing a procedure or bulleted list with the list. Keep at least two steps or list entries at the bottom or top of a page. A step's explanatory paragraph should accompany the step. Try to keep all steps in a procedure on one verso-recto page spread; avoid continuing a procedure from a recto to a verso page.
- Avoid breaking a table across pages, especially from a recto to a verso page. If breaking is unavoidable, leave the title (if applicable) and at least two rows of the table at the top or bottom of a page. Repeat the table title—followed by *continued,* all lowercase, in parentheses, and italic, including the parentheses—and the headings at the top of the next page. If an item is footnoted, the footnote goes at the end of the table. Try to keep a table's introductory sentence with the table.
- Try to keep art on the same page as the material it illustrates. Always keep an introductory phrase with its art.
- Try to have at least five or six lines on the last page of a chapter.
- In printed indexes, if the main entry is followed by subentries, do not leave the main entry alone at the bottom of the column.
- In printed indexes, if you must break up a list of subentries, at the top of the next column include the main entry followed by *continued,* all lowercase, in parentheses, and italic, including the parentheses.

palette

A collection of colors or patterns that users can apply to objects, such as the color display in Control Panel.

Users click an option from a palette. The palette name should be initial cap and bold.

Correct

Click the color of your choice from the Color palette.

pane

Use only to refer to the separate areas of a split or single window. For example, in Windows Explorer, the names of all the folders can appear in one pane and the contents of a selected folder in the other pane. Use lowercase for pane names, as in "the annotation pane."

panorama

It is acceptable to use *panorama* or *panoramic view* to describe the Microsoft Surround Video technology used in Expedia and other programs. However, if the view is full circle, use instead *360-degree* or *360° view*. Use of the degree symbol is acceptable, but note that it may be difficult to see online.

NOTE Surround Video, despite its name, is technically not a video presentation.

Parallelism

Because parallelism provides clarity and rhythm in writing, be sure that elements of sentences that are equal in purpose are also equal in grammatical structure.

Parallelism in lists

Items in lists should be parallel to each other in structure.

Correct

There are several ways to open documents in Windows. You can:

- Open your document from within the program you used to create it.
- Use the Documents command on the Start menu.
- Use the Find command on the Start menu to locate the document and then open it.
- Double-click a document icon in My Computer.

Incorrect

There are several ways to open documents in Windows:

- You can open your document from within the program you used to create it.
- Use the Documents command on the Start menu.
- The Find command on the Start menu locates the document and you can then open it.
- Double-clicking a document icon in My Computer opens a document.

Parallelism in procedures

In procedures, steps should be written in parallel style.

Correct

To share your printer

1. Click the Start button, point to Settings, and then click Printers.
2. In the Printers window, click the printer you want to share.
3. On the File menu, click Sharing.

Incorrect

To share your printer

1. Clicking the Start button, you point to Settings, and then click Printers.
2. In the Printers window, the printer you want to share should be selected.
3. On the File menu, click Sharing.

Parallelism in sentences

For parallel structure, balance parts of a sentence with their correlating parts (nouns with nouns, prepositional phrases with prepositional phrases, and so on). Sometimes, to make the parallelism clear, you may need to repeat a preposition, an article *(a, an, the),* the *to* in an infinitive, or the introductory word in a clause or phrase.

Correct

The *User's Guide* contains common tasks, visual overviews, a catalog of features, and an illustrated glossary of terms. [parallel objects, with the articles added]
With this feature you can choose which components to install and which ones to file away for later use. [parallel clauses]
Other indicators can appear on the taskbar, such as a printer representing your print job or a battery representing power on your portable computer. [parallel phrases]

Incorrect

The *User's Guide* contains common tasks, visual overviews, a catalog of features, and illustrated glossary of terms.
With this feature you can choose which components to install and the ones to file away for later use.
Other indicators can appear on the taskbar, such as a printer to represent your print job or a battery representing power on your portable computer.

parameter

Technical term referring to a value given to a variable until an operation is completed. Do not use *parameter* to mean "characteristic," "element," "limit," or "boundary."

SEE ALSO **argument**

A B C D E F G H I J K L M N O P Q R S T U V W X Y Z

parent/child

Okay to use in technical documentation to refer to the relationship in a multitasking environment or nodes in a tree structure.

Do not use as a synonym for a master/slave relationship. These terms do not mean the same thing. See **master/slave**.

parenthesis (s), parentheses (pl)

Use the term *opening parenthesis* or *closing parenthesis* for an individual parenthesis, not *open parenthesis, close parenthesis, beginning parenthesis, ending parenthesis, left parenthesis,* or *right parenthesis.* It's okay to use just *parenthesis* if it's understood which one.

In general, parentheses should be in the font style of the context of a sentence, not in the format of the text within the parentheses. For example, the text within parentheses might be italic, but the parentheses themselves would be roman if the surrounding text is roman. An exception to this is *"(continued),"* which is used for tables that continue on the next page or index subentries that continue in the next column or on the next page.

Correct

For a single-column array, use INDEX (*array,row_num*).

Passive Voice

In general, avoid the passive voice except to avoid assigning blame to a user or when the actor is unknown or immaterial. For more information, see **Active Voice vs. Passive Voice**.

path

Use *path,* not *pathname,* both in general reference and in syntax. The path describes the route the operating system follows from the root through the hierarchical structure to locate a folder, directory, or file.

To indicate a path, type first the drive name, followed by a colon and a backslash, then the name of each folder, in the order you would open them, separated by a backslash. For example:

```
C:\My Computer\Working Files\Document
```

Use *address* or *URL,* not *path,* to refer to a location on the Internet.

In general, use "path of" to refer to the location of a file.

Correct

The path of my current tax form is C:\My Computer\Working Files\Taxes\This year's taxes.

In command syntax, *path* represents only the directory portion of the full path. For example:

copy [*drive:*][*path*]*filename*

In Macintosh documentation, use colons with no spaces to separate zones, file servers, shared disks, folders, and file names. File names have no extensions.

Correct [Macintosh]

Macintosh HD:My Documentation:Sales
CORP-16:TOMCAT:EX130D Mac Temp:Workbook1

For information about capitalization of paths, see **Capitalization** and **Document Conventions**.

SEE ALSO **directory, folder, Macintosh**

PC, PC-compatible

PC is the acronym for *personal computer.* Avoid the acronym; use *computer* or *personal computer* instead. Also, because the term *PC-compatible* is vague, try to avoid it, specifying non-IBM operating environments as necessary (for example, Macintosh or UNIX). For more information about references to computer hardware, see **System Requirements**.

PC Card vs. PCMCIA

Use *PC Card,* not *PCMCIA* or *PCMCIA card,* to refer to the add-in memory and communications cards for portable computers.

p-code

Abbreviation for *pseudocode.* At first use, spell out the term and use the abbreviation in parentheses. Capitalize as *P-Code* in titles and as *P-code* when it's the first word in a sentence. Use only in programming documentation.

pen

An input device that consists of a pen-shaped stylus that interacts with a computer. Use *input device* when referring generically to pens, trackballs, styluses, and so on.

Use *tap* (and *double-tap*) instead of *click* when documenting procedures specific to pen pointing devices. *Tap* means to press the screen and then lift the pen tip.

per

In the meaning of "for each," *per* is acceptable in statistical or technical contexts. In casual or colloquial contexts, however, use *a* or *for each* instead of *per.*

Correct

Users who log on only once a day are rare.
You can have only one drive letter per network resource.

Incorrect

Users who log on only once per day are rare.

Do not use *per* to mean *by* or *in accordance with.*

A B C D E F G H I J K L M N O P Q R S T U V W X Y Z

Correct

Find all the topics that contain a specific word by following the instructions on your screen.
Identify your computer by using the procedure in the next section.

Incorrect

Find all the topics that contain a specific word, per the instructions on your screen.
Identify your computer per the procedure in the next section.

percent, percentage

One word. In general, spell out; do not use the percent sign (%), except in tables and as a technical symbol. When spelling out *percent,* put a space between the number and the word. Always use a numeral with *percent,* no matter how small.

Correct

At least 50 percent of your system resources should be available.
Only 1 percent of the test group was unable to complete the task.

Incorrect

At least 50% of your system resources should be available.
At least 50 per cent of your system resources should be available.
Only one percent of the test group was unable to complete the task.

When describing an unspecified quantity, use *percentage,* as in "a large percentage of system resources," unless doing so would be inconsistent with the interface. For example, Microsoft Project has a Percent (%) Complete field.

Periods

The following guidelines provide specific directions for using periods:

- Use only one space after a period in both printed and online documentation.
- When a colon introduces a bulleted list, use a period after each entry if it is a complete sentence or a phrase that completes the introduction. Do not end the entries with periods if they are all short phrases (three words or fewer). For more information, see **Lists**.
- Set periods in the character formatting (roman, bold, or italic) of the preceding word. If the preceding word is a command, option, keyword, placeholder, part of a code sample, or user input that requires other formatting, it is preferable to use roman for the punctuation to avoid confusion. This may cause wrapping problems online, however, so try to avoid the problem by rewriting. But use your discretion and best judgment.
- When referring to a type of file name extension, precede it with a period, as in ".prd extension" or "an .exe file." For more information, see **File Names and Extensions**.
- In numbered procedures, do not put periods after the numbers preceding each step of the procedure unless your particular design calls for them.

For more information about the use of periods, see *The Chicago Manual of Style*. For information about using the word *period* as a special character key name, see **Key Names**.

peripheral (adj)

Avoid as a noun, especially in end-user documentation. It's jargon. Use *peripheral device* or a more specific term instead.

permissions

Permissions are rules associated with a resource shared on a network, such as a file, directory, or printer; permissions can be assigned to groups, global groups, and even individual users. Be sure to distinguish permissions from *rights,* which apply to the system as a whole. Permissions and rights are "granted." Do not use *privileges* or *permission records.*

Correct

Folder designers can define specific permissions for each user.
You have permission to read, create, and edit items.

ping, PING

Do not use *ping* to refer generally to searching for a program. It's slang. It is acceptable when it refers specifically to using the PING protocol. The PING protocol is used specifically for checking the presence of a host on the Internet. It stands for *Packet Internet Groper*, but do not spell out. Describe if necessary.

pipe (n)

The symbol for a pipe in programming is a vertical bar (|). Avoid as a verb, especially in end-user documents. It's slang, except to specifically refer to routing data from the standard output of one process to the standard input of another. Instead, use a more specific term, such as *send, move, copy, direct, redirect,* or *write.*

pixel

Short for *picture element.* One pixel is a measurement representing the smallest amount of information displayed graphically on the screen as a single dot. In end-user documentation, define pixel at first use.

placeholder (n)

Do not use as a verb. For formatting of placeholders, see **Document Conventions**.

plaintext vs. plain text

Use *plaintext* only to refer to nonencrypted or decrypted text in material about encryption. Use *plain text* to refer to ASCII files.

platform

Refers to hardware architecture and is sometimes used interchangeably with *operating environment* or *environment.* But because it can be ambiguous, avoid *platform,* particularly in end-user documentation.

Platform can be used in programmer documentation if necessary to distinguish differing behaviors of a function or other API element in various operating systems, but whenever possible use *operating system* for clarity. *Cross-platform* is acceptable to refer to a program or device that can run on more than one operating system.

Use *on* to refer to a hardware platform: "on the Macintosh," but "in Windows 95."

SEE ALSO **operating environment, operating system**

Plug and Play (n, adj)

Spell as shown in all instances. That is, use title caps and do not hyphenate. Refers to a set of specifications developed by Intel for automatic configuration of a computer to work with various peripheral devices. For more information, see the *Microsoft Press Computer Dictionary.*

plug-in

Although generically *plug-in* can refer to any small program that "plugs in" to another to add functionality, in Internet usage it almost always refers to a Netscape-specific component. Do not use as a synonym for **add-in** or **add-on**.

Plurals

If you are not sure how to form the plural of a word, check the *American Heritage Dictionary.* Follow these guidelines for other plurals.

- Form the plural of an acronym by adding an *s* with no apostrophe.

  Correct
  APIs
  CPUs
  DBMSs
  VBXs

- Form the plural of a single letter by adding an apostrophe and an *s*. The letter itself (but not the *s*) is italic.

  Correct
  x's

- Form the plural of a number by adding an *s* with no apostrophe.

  Correct
  486s
  1950s

SEE ALSO **Abbreviations and Acronyms, Possessives**

point (v), point to

Use *point to* in procedures involving submenus that don't need to be clicked—for example, "Click the Windows Start button, point to Programs, and then click Windows Explorer."

Also use *point to* to mean positioning the mouse pointer at the appropriate location on the screen—for example, "point to the window border." *Move the mouse pointer to* is acceptable phrasing only when teaching beginning skills.

pointer

Mouse term. Although the mouse pointer can assume many shapes, avoid descriptive labels for mouse pointers. For example, use *pointer,* not *double-headed arrow,* unless necessary to distinguish the different types of mouse pointers. If a description would be unwieldy, use a graphic if possible.

Use *input device* generically to refer to the mouse, a pen, ball, stylus, or other input device.

For more information about pointer shapes, see *The Windows Interface Guidelines for Software Design.*

pop-up (adj)

Do not use as a noun. Also, avoid as a verb; instead, use a term that more accurately describes the message, such as *opens* or *appears.*

Use *pop-up menu* only in technical documentation. Use *shortcut menu* in end-user documentation.

Pop-up window is acceptable in references to context-sensitive Help, as in "If you want to print the information in a pop-up window, use the right mouse button to click inside it, and then click Print Topic." Do not use *pop-up window* as a synonym for *dialog box.*

Correct
Answer the questions in the wizard as they appear.
Some commands carry out an action immediately; others open a dialog box so that you can select options.
A pop-up window gives additional information about an option.

Incorrect
Answer the questions in the wizard as they pop up.
Some commands carry out an action immediately; others open a pop-up window so that you can select options.

Use "pop-up list" in Macintosh documentation to refer to unnamed list boxes.

Correct
In the pop-up list, click Microsoft Excel.

port (n)

As in *printer port* or *communications port.* Use the verb forms *port to* and *port for* only in technical documentation in reference to portability. Avoid them in end-user documentation.

A B C D E F G H I J K L M N O P Q R S T U V W X Y Z

portable computer

Whenever possible, use the generic term *portable computer* in discussions about notebook, laptop, and other portable computers. In general, avoid these other designations because each refers specifically to a particular size and weight range. For more information about sizes, see *Microsoft Press Computer Dictionary.*

Do not use *portable* as a noun.

portrait orientation

Printing orientation that prints across the narrow side of the paper.

Portrait orientation

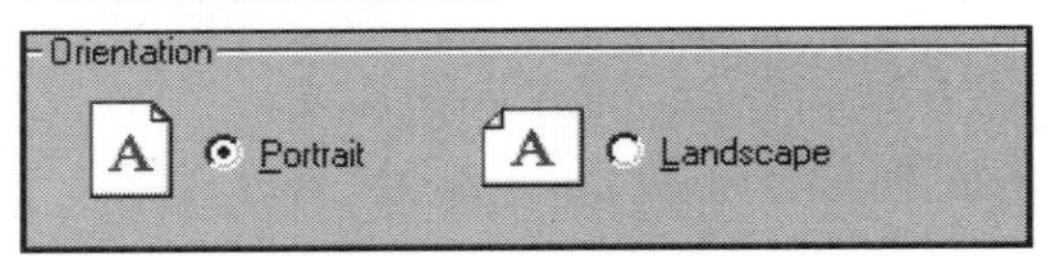

Compare **landscape orientation**.

Possessives

Form the possessive of singular nouns and acronyms by adding an apostrophe and an *s*. Form the possessive of plural nouns that end in *s* by adding only an apostrophe. Form the possessive of plural nouns that do not end in *s* by adding an apostrophe and an *s*.

Correct

the encyclopedia's search capabilities
a children's encyclopedia
an OEM's products
the articles' links

It is acceptable to form the possessive of acronyms, but avoid it if possible if the acronym does not refer to people or companies. Either use the name with no ownership or use an *of* phrase or a similar rewrite. It is acceptable to form the possessive of company names.

Do not use possessives for product or feature names.

Correct

the Windows interface
Word templates, the templates in Word
the dictionary in the spelling checker
the Send command on the File menu
the products of OEMs
Microsoft's benefits program

Incorrect

Windows's interface
Word's templates
the spelling checker's dictionary
the File menu's Send command

SEE ALSO **Apostrophes**

post (v)

Use *post* to refer to the act of sending a message to a newsgroup or public folder, as opposed to sending it to a person. Avoid using *post* as a noun to refer to the item sent, unless necessary for consistency with the interface. (The use of *post* as both a noun and a verb in commands under one menu can be confusing.) Instead, use *article, message*, or other specific term, depending on the context, to refer to the material sent.

Avoid using *post* as a synonym for *publish,* especially when referring to publishing material on the Web.

post office vs. postoffice

One word, lowercase, when referring to the component of a Microsoft Mail system; otherwise, use two words.

pound key, pound sign (#)

Do not use either term to refer to the keyboard key name; use *number sign* instead. It is acceptable, however, to use *pound key (#)* when referring specifically to telephones or the telephone keypad.

power cord

Not *power cable.*

power down, power up; power off, power on

Do not use; use *turn off* and *turn on* instead. Do not use *shut down* to refer to turning off a computer.

SEE ALSO **shut down, shutdown, Shut Down, turn on, turn off**

pre (prefix)

In general, do not hyphenate words beginning with *pre,* such as *preallocate* and *preempt,* unless it's necessary to avoid confusion, as in *pre-engineered,* or if *pre* is followed by a proper noun, as in *pre-C++*. If in doubt, check *American Heritage Dictionary* or your project style sheet.

preceding

Use *preceding* or *earlier* to mean earlier in a book or Help topic instead of *above.*

SEE ALSO **Cross-References**

A B C D E F G H I J K L M N O P Q R S T U V W X Y Z

Preface

Do not use "Preface" as the title of the introductory material in documentation. Use "Introduction" or a more descriptive title appropriate to the audience, such as "Before You Begin."

Preferred Method

In user assistance materials, always document the preferred method of performing an action or procedure, even though more methods may be possible. This preferred method is generally determined by the product team, keeping in mind the needs of the audience.

Occasionally, alternative methods are provided in addition to the preferred method.

SEE ALSO **Procedures**

Prefixes

In general, do not use a hyphen between a prefix and a stem word unless a confusing word would result or if the stem word begins with a capital letter. For more information, see the individual entries in this guide for specific prefixes, *American Heritage Dictionary,* or *The Chicago Manual of Style.*

Prepositions

Ending a sentence with a preposition is acceptable to prevent an awkward construction, but avoid stacking prepositional phrases on top of one another.

Correct

Type the text you want to search for.
After you click Save As on the File menu, Word displays a dialog box. In the lower-right corner, click Options.

Incorrect

Type the text for which you want to search.
The Options button is in the rightmost corner of the dialog box of the Save As command on the File menu.

For more information, see the individual entries in this guide for specific prepositions.

press

Differentiate among the terms *press, type, enter,* and *use.* Use the following guidelines:

- Use *press,* not *depress* or *type,* when pressing a key initiates an action within the program or moves the user's position within a document or worksheet—for example, "press ENTER" or "press N." Use *pressed in* and *not pressed in,* not *depressed* and *not depressed,* to refer to the position of 3-D toggle keys.
- Use *use* in situations when *press* might be confusing, such as when referring to a type of key such as the arrow keys or function keys. In such cases, *press* might make users think they need to press all the keys simultaneously—for example, "use the arrow keys to move around the document."

- Use *type,* not *enter,* to direct a user to type information that will appear on the screen—for example, "type your name."
- Do not use *strike* or *hit.*
- Do not use *press* as a synonym for *click.*

Correct
Type your name, and then press ENTER.
Press CTRL+F, and then type the text you want to search for.
To save your file, press Y.
To move the insertion point, use the arrow keys.

Incorrect
To save your file, use CTRL+S.
Hit ENTER to begin a new paragraph.

Press

It is acceptable to use *Press* to refer to *Microsoft Press,* if it's not done too often and if the full name is used frequently. Do not shorten to *the Press* or *MS Press.*

print (v), printout (n)

Use *print,* not *print out,* as a verb. It's all right to use *printout* as the result of a print job, if necessary, but try to be more specific.

print queue

Not *printer queue.*

privileges

Do not use as a synonym for *permissions* or *rights.* See **permissions** and **rights**.

Procedures

In documentation, a procedure is a short description of the steps a user takes to complete a specific task. In printed and online documentation, procedures are set off from the main text by their formatting. In Help, a procedure may be on a separate "how-to" screen, as well.

In most materials, always document the preferred method of performing a procedure if there is more than one way to do something. The preferred method should reflect the needs of the audience as much as possible. Note, however, that the mouse method may cause difficulties for users with some disabilities. One way to present alternative methods is by judiciously using tips. For information about documenting an alternative method in addition to the preferred method, see **Branching within procedures**, on page 220.

Always present a procedure (except a single-step procedure) in a numbered list and in most cases introduce it with an infinitive phrase. The procedure heading serves as the introductory phrase, which follows the style of your design template. Do not add punctuation at the end of the heading.

This is a typical procedure heading:

To merge subdocuments

Avoid intervening text between the procedure heading and the numbered list.

Procedure lists

Most procedures consist of a number of steps. Try to limit a procedure to seven or fewer steps. Instructional design experts say this is the maximum number of items people can remember at once.

Multiple-step procedures

General rules for lists also apply to procedure lists—especially the following:

- Set individual steps as separate, numbered entries. However, short steps, if they occur in the same place (within one dialog box, for example), can be combined.

  Both of the following examples are correct, although the first is more commonly used.

  Correct

  1. On the Tools menu, click Options, and then click the Edit tab.
  2. …

  1. On the Tools menu, click Options.
  2. Click the Edit tab.

- Do not use a period following the step number unless your design requires it.
- Use complete sentences.
- Use parallel construction.
- Capitalize the first word in each step.
- Use a period after each step. An exception is when you are instructing users to type input that does not include end punctuation. In this case, try to format the text so the user input appears on a new line.
- In printed documentation, try to keep all steps in a procedure on one page or left-right (verso-recto) page spread, and avoid continuing a procedure across a right-left (recto-verso) break. Online, keep a procedure to one screen.
- Avoid burying procedural information in narrative text; the procedure will be hard to find and follow.

Single-step procedures

Most designs have a single-step bullet to mark a single-step procedure. Each design specifies the type of single-step bullet. Never number a single-step procedure as “1.”

Correct

To look at the PERT Chart

- On the View menu, click PERT Chart.

Incorrect

To look at the PERT Chart

1. On the View menu, click PERT Chart.

Procedural syntax

As a general rule, tell the user where the action should take place before describing the action to take. This prevents users from doing the right thing in the wrong place. However, avoid overloading procedures with locators. Assume that the user is looking at the screen and is starting from the position where the procedure begins. For example, the following phrasing is typical: "On the View menu, click Zoom."

However, if there's a chance users might be confused about where the action should take place or if an introductory phrase is needed, the following wording can be used: "If you want to magnify your document, click View, and then click Zoom," or "In Control Panel, double-click Passwords, and then click Change Passwords."

NOTE It's not necessary to end a procedure with "Click OK" unless there's some possibility of confusion.

The following sections give brief guidelines on how to treat the main elements in the interface: folders and icons, commands, and dialog box buttons.

Folders and icons

Users *click* or *double-click* a folder or an icon to initiate an action—for example, starting a program or viewing a list of subfolders.

When you want the user to	*Use this syntax*
Activate a program icon that is already running on the desktop	Click the Microsoft Excel button on the taskbar. Switch to Microsoft Excel.
Start a program	Click the Start button, and then point to Programs. Point to the folder that contains Word, and then click Word.
Select an icon before changing its properties, moving it, and so on	Right-click the PowerPoint icon, and then click Properties.
Choose a Control Panel icon	In Control Panel, double-click the Printers icon.
Choose any other icon, such as a folder icon, drive icon, and so on	Double-click the Recycle Bin icon.

A B C D E F G H I J K L M N O P Q R S T U V W X Y Z

Commands

Use the following syntax for commands. If your group decides to document more than just mouse procedures, use the word *choose* instead of *click* and use *from the menu,* not *on the menu.*

When you want the user to	*Use this syntax*
Carry out a command from a program menu	On the Tools menu, click Address Book.
Carry out an action from a dialog box reached from a menu command	On the Tools menu, point to Language, click Hyphenation, and then select the Automatically hyphenate document check box.
Carry out a command from a dialog box	Click Apply.
Carry out a command from a submenu	Click the Start button, point to Documents, and then click the document you want. Click the Start button, point to Programs, and then click Windows Explorer.

Dialog box options

Use *click* for selecting dialog box options and tabs and for choosing command buttons. However, if your group uses the combined mouse and keyboard method of documenting procedures, use *choose* for command buttons, because command buttons actually carry out a command and are not options (or settings) in the sense that the other elements of a dialog box are. In this method, use *select* for all other dialog box options, including tabs.

Tell users to *type* in text boxes or *enter* an element such as a file name that they can either type or select from a list. When you are referring generally to a feature, *turn off* and *turn on* are acceptable. Use *select* and *clear* to refer to check boxes. You can say "click to select" if the action may not be obvious to your audience.

The following table gives a few examples. For procedural syntax for every dialog box and option, see **Dialog Boxes and Property Sheets**.

When you want the user to	*Use this syntax*
Select an option	In the Print dialog box, click All.
Open a tabbed section in a dialog box	In the Font dialog box, click the Character Spacing tab.
Insert something in a text box that has a list box attached	In the File name box, enter the name of the file.
Select among check boxes	On the Print tab, select the Comments and Hidden text check boxes.
Look in a particular grouping of options or check boxes	Under Include with document, select the Comments and Hidden text check boxes.

Procedure style

Follow standard document conventions in procedures.

- Follow interface capitalization. Usually, menu and command names use title caps. Capitalization of dialog box options varies. If in doubt, or if necessary for consistency, use sentence style caps.

  Correct
  Click Date and Time.
  Select the Provide feedback with sound check box.

- If a command name or dialog box option ends with a colon or ellipsis, do not include this punctuation.

  Correct
  Click Save As.

  Incorrect
  Click Save As ...

- Limit your use of the descriptors *button* and *option button,* except where the descriptor helps to avoid confusing or awkward phrasing or is necessary to avoid confusion with another element.
- Use bold for user input and italic for placeholders. User input can be on the same line as the procedural step, or it can be displayed on a new line. If the input is on the same line, what the user types should be the last word or words of the step and should not be followed by end punctuation (unless the user needs to type the end punctuation).

  Correct
  Type *password*
  In the Date box, type:
  April 1

- Use a monospace font for program input and output text.

SEE ALSO **Code Formatting Conventions, Dialog Boxes and Property Sheets, Document Conventions, Menus and Commands**

Mouse vs. keyboard procedures

You can document procedures in one of three ways:

- Mouse-only actions, using terms such as *click, double-click,* and *point to.*

  NOTE For pen-computing documentation, use the words *tap* and *double-tap.*

- Combined mouse and keyboard actions, using general procedural terms such as *select* and *choose* to avoid distinguishing between the two methods.

  NOTE In training sessions or when the user is likely to be a novice, it may be appropriate to provide specific instructions for each. In this case, use phrases such as "click OK or press ENTER

- Separate mouse and keyboard actions, explaining first the mouse method and then the keyboard method. Whenever possible, avoid a page break between the keyboard and the mouse versions of a procedure.

The trend is to document with mouse-only actions. Note, however, that this method makes documentation less accessible to people with certain disabilities. The method of documenting procedures is generally a group or lead decision.

A B C D E F G H I J K L M N O P Q R S T U V W X Y Z

Branching within procedures

If there are multiple ways to do an entire procedure—and if you have to describe each alternative—use a table to detail the alternatives, similar to the following example. This helps the user know when to use which method.

Correct

To	*Do this*
Save changes to the existing file	On the File menu, click Save.
Quit without saving changes	On the File menu, click Exit.

If one step has an alternative, that alternative should be a separate paragraph in the step. In a single-step procedure, an alternative can be separated by the word *or* to make it clearer to the user that an alternative is available.

Correct

1. Press the key for the underlined letter in the menu name.
   You can also use the arrow keys to choose another menu.

To open a menu

- Press ALT+the key for the underlined letter in the menu name.
  – or –
  Use the arrow keys to choose another menu.

Incorrect

Press ALT+the key for the underlined letter in the menu name. You can also use the arrow keys to choose another menu.

1. Press the key for the underlined letter in the menu name; or,

1. Use the arrow keys to choose another menu.

For several choices within one procedure step, use a bulleted list.

Correct

1. Select the text you want to move or copy.
2. Do one of the following:
   - To move the selection, click the Cut button on the toolbar.
   - To copy the selection, click the Copy button on the toolbar.
3. Position the insertion point in the new location, and then click Paste.

Supplementary information and text within procedures

Avoid putting supplementary information, which describes special cases or behaviors that are not essential to completing the procedure, within the procedure itself. If supplementary information is necessary to describe a procedure, put it in a single paragraph after the procedure or the step it explains and indent it to align with the procedure text. If supplementary information consists of steps, make it a separate procedure.

Within procedures, use commentary text sparingly. The comment should relate to the procedure step and should contain information the user needs to move on to the next step. Put commentary text in a separate paragraph following the step and indent it to align with the procedure text.

Correct

1. Click Options.

   The dialog box expands to display more options.
2. Click to select the Reverse Print Order check box.

For information about documenting key commands, see **Key Names** and **press**.

program file

Okay to use if necessary, especially to avoid *executable file* in end-user documentation, but use the specific name of the file whenever possible.

program vs. application

If possible, refer to a product by its descriptor, such as *database management system, spreadsheet,* or *publishing toolkit* rather than *program* or *application.* For example, refer to Microsoft Visual FoxPro as "the Microsoft Visual FoxPro relational database development system."

If that's not possible, follow these general guidelines.

- *Program* is considered the friendlier term, so use *program,* not *application,* in material written for end users. It usually refers to an executable file.
- Use *application* in material whose audience is developers, administrators, and those who provide technical support, especially to refer to a grouping of software that includes both executable files and other software components, such as a database.
- Do not use *program application.*

If in doubt, consult your project's style sheet.

SEE ALSO **applet**

A B C D E F G H I J K L M N O P Q R S T U V W X Y Z

progress indicator

A control that displays the percentage of a particular process that has been completed, such as printing or setting up a program. Do not refer to it as a "slider."

Progress indicator

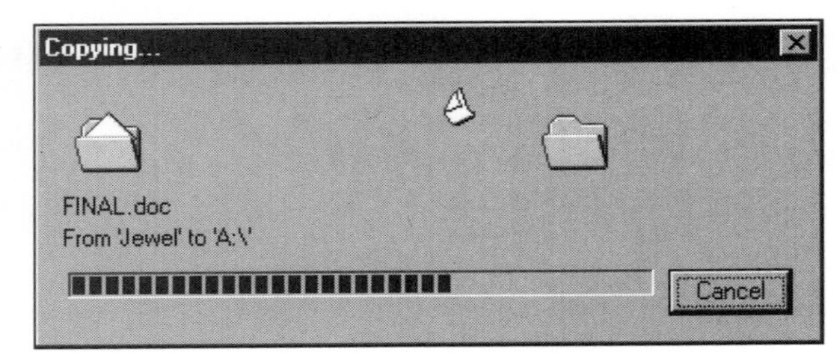

Project

Always use *Microsoft Project;* do not shorten to *Project.*

prompt (v)

Do not use *prompt* as a noun to mean "message." Use *prompt* as a verb to mean the system is requesting information or an action from the user.

Correct

If you receive a message that the association information is incomplete ...
When you run Setup, you are prompted to insert disks one by one.

Incorrect

If you receive a prompt that the association information is incomplete ...

SEE ALSO **command prompt**

prop (v)

Do not use in documentation to refer to propagating files to a server or information to a database. It's slang.

properties

Properties are attributes or characteristics of an object used to define its state, appearance, or value. For example, the Special Effect property in Microsoft Access determines the appearance of a control on a form, such as sunken, raised, flat, and so on.

However, because the term can be vague, avoid it except for a specific reference to something named as a property. Use instead *value* or *setting* to refer to a specific characteristic a user can set (such as the specific color of a font) or *attribute* for the general characteristic (such as "color is an attribute of fonts").

In syntax, properties are often bold, but always check your project style sheet.

property sheet, property page

Property sheet refers either to a secondary window that displays the properties of an object after carrying out the Properties command or to a collection of tabs or *property pages* that make up a dialog box.

In general, do not use the terms *property sheet* and *property page* in end-user material; use *tab* or *dialog box* instead. If your product uses *property sheets,* see your project style sheet for specific usage of the term.

SEE ALSO **Dialog Boxes and Property Sheets**

protected mode

Not *protect mode.* Use only in technical documentation.

Protocols

A protocol is a set of rules or standards that allow computers to communicate. Most protocols are referred to by their abbreviations. For example, SMTP is an abbreviation for Simple Mail Transfer Protocol.

In Internet addresses, the protocol used by the Web server appears (in lowercase) before a colon; it specifies the access scheme. Typical Web protocols are HTTP, FTP, news, and so on.

Use initial caps for protocol names (except in URLs) unless you know the name is handled differently. If in doubt, check the index to this guide or follow your project style sheet.

pull-down (adj)

Do not use as a noun. Acceptable as a descriptor for Macintosh menus.

SEE ALSO **drop-down**

pull quote

Two words. Refers to an excerpt from the main text printed as a sidebar or other notation to get the reader's attention.

pull, push

These terms refer to technology in which information can automatically be sent to a user via Web pages, software components, e-mail messages, or other broadcast content. "Push technology" describes an interaction where the server automatically uploads data without a specific command from the client. In "pull technology," the client configures the product to pull in new information only on a specified schedule or only specified information.

A B C D E F G H I J K L M N O P Q R S T U V W X Y Z

punctuation

In general, set colons, commas, periods, and semicolons in the character formatting (roman, bold, or italic) of the preceding word unless the preceding word is a command, option, keyword, placeholder, part of a code sample, or user input. In these cases, it is preferable to use roman formatting. This may cause wrapping problems online, however, so try to avoid the problem by rewriting. But use your discretion and best judgment.

Set quotation marks, exclamation points, and question marks in the character formatting of the word preceding them if they are part of that word:

Correct

Type **Balance Due:** in cell A14.

In online documentation, use standard paragraph formatting for punctuation marks following a jump or pop-up text.

When parentheses or brackets appear within a sentence, set them in the formatting of the text outside the marks, not the text within the marks. An exception to this is *(continued),* which is used for table headings and in indexes. Never use two different styles, such as italic for an opening parenthesis and roman for a closing parenthesis. Use the predominant sentence formatting.

For information about specific punctuation marks, see the individual entries in this guide and in *Harbrace College Handbook* and *The Chicago Manual of Style.*

push button (n), push-button (adj)

Avoid in all contexts; instead, use *command button* to refer to the control. In technical material, however, *push button* can be included parenthetically and in indexes, if necessary. Do not spell as one word (that is, not *pushbutton*).

In end-user material, just use the specific name of the button (as in "click Apply"). Avoid uses such as "the Apply button," and do not use "the Apply command button" or "the Apply push button."

quality (n)

Do not use as an adjective.

Correct

Microsoft Word has a number of high-powered qualities.
Microsoft Word is a high-quality word processor.

Incorrect

Microsoft Word is a quality product.

SEE ALSO **high-quality**

quarter inch

In general, use *a quarter inch,* not *quarter of an inch* or *one-quarter inch.*

quick key

Do not use; instead, use *shortcut key* or *access key,* depending on meaning.

SEE ALSO **access key, Key Names, shortcut key**

QuickBasic

Always use *Microsoft QuickBasic;* do not shorten to *QuickBasic.*

quit (a program)

Use *quit* instead of *close, exit, end, leave, stop,* or *terminate* to refer to closing a program. Note that the user can click the Close button or the Exit command to quit, but use *close* to refer to either the button or closing a window and *exit* only to refer to the command. Use *log off* only to refer to ending a network or Internet connection.

Quotation Marks

Not *quote marks* or *quotes.*

In print documentation use curly quotation marks (“ ”), sometimes called “smart quotes,” for everything except user input and code samples, which call for straight quotation marks ("). The term *quotation marks,* unless qualified, refers to double curly quotation marks.

In online documentation, use straight quotation marks. They are automatically formatted to be straight.

Use the terms *opening quotation marks* or *closing quotation marks,* not *open quotation marks, close quotation marks, beginning quotation marks,* or *ending quotation marks.*

Placement of quotation marks

Conventions of American English call for quotation marks to appear outside commas and periods. Their placement with question marks and semicolons depends on whether the punctuation mark is part of the material being quoted. However, quotation marks have specialized uses in many computer languages. Follow the conventions of the language in code samples.

Do not put quotation marks inside periods and commas in normal textual material.

Correct

One Internet dictionary calls an electronic magazine a “hyperzine.”
A “hyperzine,” according to one Internet dictionary, is an electronic magazine.
The Web lists a number of new “hyperzines”; *Slate* is a new one from Microsoft.

```
/*Declare the string to have length of "constant+1".*/
```

Incorrect

One Internet dictionary calls an electronic magazine a “hyperzine”.
A “hyperzine”, according to one Internet dictionary, is an electronic magazine.
The Web lists a number of new “hyperzines;” *Slate* is a new one from Microsoft.

radio button

Refers to *option button,* the round buttons in dialog boxes that users select to make one choice from among several items. Do not use in end-user documentation; use *option* or the specific name of the option instead. Avoid in technical documentation. If necessary for a particular technical audience or to maintain consistency with previous documentation, use *option button,* noting on first use that it's also called *radio button,* and include in the index as a synonym for *option button.*

SEE ALSO **Dialog Boxes and Property Sheets, option, option button**

radix (s), radixes (pl)

Do not use *radices* as the plural.

ragged right

Okay to use to refer to the uneven right edge in documents. Opposite of *right-aligned.*

RAM

Acronym for *random access memory.* Spell out at first mention, unless you are positive that your audience knows the term. See **memory**.

range selection

In end-user documentation, use a phrase such as "a range of items," rather than *range selection,* to refer to a selection of adjoining pages, cells, and so on. Use the same type of phrasing in technical documentation, but if you are describing the feature use *adjacent selection. Contiguous selection* is acceptable if the term is used elsewhere in the product, but try to avoid.

The selection of more than one nonadjacent item is called a *multiple selection.*

SEE ALSO **adjacent selection, multiple selection**

A B C D E F G H I J K L M N O P Q R S T U V W X Y Z

re (prefix)

In general, do not hyphenate words beginning with *re* unless it's necessary to avoid confusion or *re* is followed by a word that is ordinarily capitalized. If in doubt, check *American Heritage Dictionary* or your project style sheet.

Correct

reenter
recover [to get back or regain] versus re-cover [to cover again]
recreate [to take part in a recreational activity] versus re-create [to create anew]

Also, avoid using *re* words (such as *resize* or *reboot*) unless you mean that a user should redo or repeat an action. The root word *(size* or *boot)* is often enough.

Readme Files and Release Notes

Readme files and release notes often contain similar types of information and can usually be treated in the same way. The main difference is that readme files provide up-to-the-minute information about a newly released product, and release notes provide information about test and beta releases.

You can use the term *readme file* or *readme* without an extension, but if it is a text file the extension *.txt* can clarify to the user that it will appear as a text file without character formatting. Capitalize *readme* when you refer to the specific file.

Correct

Look in Readme.txt on Disk A for the most current information.
Look in the Readme file on Disk A for the most current information.

As far as practicable, the same rules of style and usage pertain to readme files as to all other documentation. Even if they don't follow the same formatting standards as documentation, readme files should not contain jargon and overly technical language and should otherwise conform to Microsoft style.

NOTE Do not use trademark symbols or notes in readme files.

Readme text files

Most readme files are either text files, formatted in Courier, or Help files. Use the following guidelines for organization, content, and formatting of readme text files. See the example following the guidelines.

Front matter

Include these elements at the beginning of the file, following the formatting guidelines listed here and in the example readme file.

- Title of the file centered in the text area, with the date (month and year) centered one line below. Insert a row of hyphens above and below.
- Standard Microsoft copyright notice, centered under the title. Use a lowercase "c" in parentheses for the copyright symbol.
- Introductory paragraph explaining the purpose of the file, flush left.
- Optional section titled "How to Use This Document," flush left, beginning two lines below the introductory paragraph and with one row of hyphens above and below. Use the boilerplate shown in the example readme file.
- Contents listing all the section headings. In general, order the readme file with the most important information or information of the most general interest first. List errata and changes to the documentation last. Section numbers, as shown in the example readme file, are optional.

Sections and topics

This section describes how to format the information in the readme file, including procedures, tables, and errata and corrections. Samples of these are included in the example readme file.

Procedures

Procedures within the text should follow the same general guidelines as all procedures. Begin with an infinitive phrase preceded by three right angle brackets (>>>) and followed by a colon. Number the steps.

Because the text file won't show character formatting, use all uppercase letters to indicate user input, use title caps for interface elements, and underline words to indicate placeholders.

Tables

You can use tables to list and describe included files and other information. Underline each row of a table heading with one row of hyphens.

Errata and Corrections

If you are listing corrections to documentation, be as specific as possible about the location and the change necessary. For example, for a book, list chapter number and chapter title, section heading, and page number. Then tell the user what to replace or add. The words *chapter, section,* and so on should be flush left, with two spaces between the longest word and the beginning of the correction text. Align as shown in the example readme file.

Help Files

For Help files, list the specific Help file and the topic or topics.

Example readme file

```
          ------------------------------------------
             Microsoft [Product Name] Readme File
                         January 1998
          ------------------------------------------
  (c) Microsoft Corporation, 1998. All rights reserved.

This document provides late-breaking or other information
that supplements the Microsoft [Product Name] documentation.
```

NOTE You can use the following paragraph where there is no other documentation:

This document provides information about Microsoft [Product Name], as well as answers to questions you might have.

```
-------------------------
How to Use This Document
-------------------------
To view the Readme file on-screen in Windows Notepad,
maximize the Notepad window. On the Edit menu, click Word
Wrap. To print the Readme file, open it in Notepad
or another word processor, and then use the Print command
on the File menu.
---------
CONTENTS
---------
1.    WHAT'S NEW IN THIS RELEASE
      1.1  One-Step Installation
      1.2  Windows Control Center

2.    INSTALLATION NOTES
      2.1  Installing over a Previous Version
      2.2  Manually Decompressing Application Files

3.    TROUBLESHOOTING
      3.1  Installation Problems
      3.2  Restoring AUTOEXEC.BAT, CONFIG.SYS, WIN.INI, and
           SYSTEM.INI
      3.3  Renaming the Installed Directory

4.    APPLICATION DISK CONTENTS
      4.1  Driver Files
      4.2  Windows Program Files
      4.3  MS-DOS Program Files
      4.4  Installation Program Files
.
.
.
```

```
>>>To change your SmartDrive cache:

1 Open WordPad.
2 On the File menu, click Open.
.
.
.
7 On the File menu, click Exit.
When the WordPad dialog box asks if you want to save your changes, click Yes.
.
.
.
Field              Description
----------         --------------------------------
Language           Three-letter language identifier
Welcome String     Text to display when the custom Setup program starts
.
.
.
Chapter:   6, "Using OLE Custom Controls"
Section:   "Using OLE Custom Control Methods"
Page:      173
           Replace "calendar control example" in the first
           sentence on the page with "calendar control example
           described earlier in this chapter" to clarify.

Help:      Scroll Bar Control Help
Topic:     Value Property, Change Event Example
           Replace the Dim statement at the beginning of the
           sbMonth_Change() procedure with the following
           statement:
           Dim Diff As Integer, i As Integer
```

Readme Help files

If you release the readme file as a Help file, follow the same general guidelines for content, but include a general Readme Help contents section on the first screen. The first section should be this boilerplate:

Microsoft [Product Name] Product Update

Late-breaking information about this release of Microsoft [Product Name].

Describe briefly the content of each Help section in the contents. Include standard instructions for printing, such as these:

To print a Readme Help topic

1. Display the topic you want in the Help window.
2. On the File menu in the Help window, click Print Topic.

A B C D E F G H I J K L M N O P Q R S T U V W X Y Z

Vary these instructions according to the setup of the Help file. For example, in Windows 95-based programs, you can organize the readme file as one "book" in Help, which can then be printed in its entirety. Alternatively, you can compile Help readme files as one long topic so that users can print the entire file.

read-only (adj)

Always spell with a hyphen regardless of its position in the sentence. Use lowercase to refer to read-only as a file attribute.

Correct

read-only memory
This file is read-only.

read/write

Use *read/write,* not *read-write,* as in "read/write permission."

read/write permission

Not *read/write access.* Files and devices have read/write properties, but users have the permission to access those files and devices.

real-time (adj)

Two words. Always hyphenate before a noun.

Correct

Real-time operations happen at the same rate as human perceptions of time.
In chat rooms users communicate in real time.

reboot

Avoid; use *restart* instead.

Recycle Bin

Use with the article *the,* as in "the Recycle Bin." In Windows 95 and later, the Recycle Bin is a temporary storage place for deleted files.

Recycle Bin

refresh

Use *refresh* to refer to updating a Web page. Avoid using in documentation to describe the action of an image being restored on the screen or data being updated; instead, use *redraw* or *update*. To refer to the Refresh command, use language such as: "To update the screen, click Refresh."

region selection

Technical term. In Windows 95 and later, a selection technique that involves dragging a bounding outline to define the selected objects.

registry, registry keys

Technical term. The *registry* is a database that stores configuration data about the user, the installed programs and applications, and the specific hardware. The registry has a hierarchical structure, with the first level in the path called a subtree. The next level is called a *key,* and all levels below keys are *subkeys.*

Use lowercase for the word *registry* except when it's part of a named system component, such as the Registry Editor. The first-level subtrees are system-defined and are in all uppercase letters, with words separated by underscores. Registry subtrees are usually bold.

Correct

HKEY_CLASSES_ROOT
HKEY_LOCAL_MACHINE

Keys are developer-defined and are usually all uppercase or mixed caps, with no underscores. Subkeys are usually mixed case.

Correct

SOFTWARE
ApplicationIdentifier
Application Identifier *Name*
stockfile
the **new program** subkey

An entire subkey path is referred to as a *subkey,* not a *path.* This is a typical subkey:

Correct

\HKEY_LOCAL_MACHINE\SOFTWARE\Microsoft\Jet\3.5\Engines\Xbase subkey

For a subkey, the items in the Name column are *entries.* The items in the Data column are *values.*

> **NOTE** This information is subject to change.

reinitialize

Do not use to mean *restart.* See **initialize**.

release note

Provides information about test and beta versions of a product. See **Readme Files and Release Notes**.

A B C D E F G H I J K L M N O P Q R S T U V W X Y Z

A B C D E F G H I J K L M N O P Q R S T U V W X Y Z

REM statement

Short for "remark statement," which is the term for a comment in BASIC and some other programs. Do not use generically to refer to a comment. Use *comment* instead.

SEE ALSO **Code Commenting Conventions**

remote (adj)

Acceptable to use to refer to a person or computer at another site, but not to a drive on a remote computer. In that case, use *network drive* instead.

Do not use as a noun except for a remote control device.

remove

Do not use *remove* to mean *delete*. *Remove* is correct, however, to refer to removing a program and items such as toolbar buttons or column headings in programs such as Outlook to customize an interface and to refer to removing a program using the Add/Remove Programs dialog box.

replace (v)

Do not use as a noun.

Correct

You can replace all instances of an Incorrect term at one time.

Incorrect

You can do a global replace of an Incorrect term.

In general, use *replace,* not *overwrite,* in instances of replacing a file.

Correct

Replace the selected text with the new text.
Replace the file with the changed file.

SEE ALSO **find and replace**

restore

Do not use the technical jargon *undelete.*

Restore button

Do not use *Restore box* or *Restore icon.* Refers to the button containing the image of two windows. It appears in the upper-right corner of a window near the Close button and can replace either the Minimize or, more often, the Maximize button. Clicking it restores a document to its previous size.

right

Not *right-hand.* Use *upper right* or *lower right, rightmost,* and so on. Include a hyphen if modifying a noun, as in *upper-right corner.*

right align (v), right-aligned (adj)

Use to refer to text that's aligned at the right margin. Hyphenate the adjective in all positions in the sentence. Do not use *right-justified.*

right mouse button

In most documentation, use this term rather than *secondary mouse button, mouse button 2,* or other terms. Even though a user can program a mouse to switch buttons, usability studies show that most users understand this commonly used term.

SEE ALSO **mouse**

right-click (v)

Acceptable to use, but define it first if necessary.

Correct

Using the right mouse button (right-click) ...
Right-click to select the file.

right-hand

Do not use; use just *right* instead.

SEE ALSO **right**

rightmost (adj)

One word. Use to refer to something at the farthest right side. Use instead of *farthest right, far-right,* or similar terms.

rights

Technical term. *Rights* are rules associated with the system as a whole, granted to local groups, global groups, and users. They allow users to perform certain actions on the system. Distinguish from *permissions,* which are associated with objects.

Correct

You have the right to log on to the network as a guest and permission to read the files.

Incorrect

You have permission to log on to the network as a guest and the right to read the files.

SEE ALSO **permissions**

ROM

Acronym for *read-only memory.* Spell out at first mention, unless you are positive that your audience knows the term. See **memory**.

roman

In general, don't use *roman type, light type,* or *lightface* unless you need to define or describe the font style; instead, use just *roman.* Do not use as a verb. Note that it is not capitalized.

Correct

Use roman, rather than italic, for most text.

SEE ALSO **Document Conventions, font and font style**

root directory

Use this term, not *home directory,* to refer to the directory or folder (indicated in MS-DOS with a backslash: \) from which all other directories or folders branch. Do not shorten to *root* when you mean the directory.

Correct

Change to the root directory and type
edit autoexec.bat

run vs. execute

Avoid using *execute* to refer to carrying out a command or running a program, especially in end-user documentation. Use *run* to refer to programs, queries, and macros and *carry out* to refer to actions and commands. In technical documentation, it's acceptable to use *execute* to refer to programming processes, especially if required by the interface, but if possible use *run* or write around the use.

Correct

While Windows defragments your disk, you can use your computer to carry out other tasks.
You can temporarily stop Disk Defragmenter so you can run other programs at full speed.

run time (n), run-time (adj)

Never run together into one word (that is, not *runtime*). Capitalize both words in headings.

runs vs. runs on

A computer runs an operating system such as Windows NT Server, but a program runs on the operating system.

Correct

Many companies are configuring their computers to run Windows NT Server.
They have to install upgraded programs to run on Windows NT Server.

running foot, running head

Use *footer* and *header* instead; *running foot* and *running head* are acceptable if needed for clarification or as keywords or index entries.

(s)

Do not add *(s)* to a singular noun to indicate that it can be singular or plural. In general, use plural, but be guided by meaning. Alternatively, if it's important to indicate both, use *one or more*.

Correct

To add rows or columns to a table, ...
To add one or more rows to a table, ...

Incorrect

To add a row(s) or columns(s) to a table, ...

save (v)

Do not use as a noun.

Correct

Before you turn off your computer, save your files.

Incorrect

Before you turn off your computer, do a save of your files.

Use these prepositions with save: "save *on* a disk," "save *to* a file," and "save *for* a rainy day."

scale up

In general, use *scale up,* not *upsize,* even though the jargon *upsize* is becoming common in client/server products.

scan line

Two words. Refers to either the row of pixels read by a scanning device or one of the horizontal lines on a display screen.

screen

Use instead of *screenful* or *full screen,* except that it's okay to refer to a program running in *full-screen mode*. Use *screen* instead of *display* or *monitor* to refer to the graphic portion of a monitor. Reserve *display* to refer to the display device (such as the kind of video adapter) and *monitor* to refer to the hardware that includes the screen.

ScreenTip

Generic term for any on-screen tip, such as a ToolTip. Used especially in end-user documentation in Office.

screen resolution

Use *number* × *number,* not *number by number,* as in "640 × 480," not "640 by 480." Use the multiplication sign, not an *x.*

Screen Terminology

The following illustrations show a typical Internet Explorer screen and Web page, a Windows 98 desktop, an open window, an open document, and a menu, with the various elements that appear on them called out. The callouts use capitalization that matches how the item name appears in documentation. Usual Microsoft capitalization style for callouts is initial cap.

Elements that may appear on some or all of the screens are not necessarily called out on all of them here, but use the same term—for example, *scroll bar* or *Close button*—regardless of the kind of screen it appears on.

For the names of all dialog box elements, see **Dialog Boxes and Property Sheets**.

Windows 98 desktop

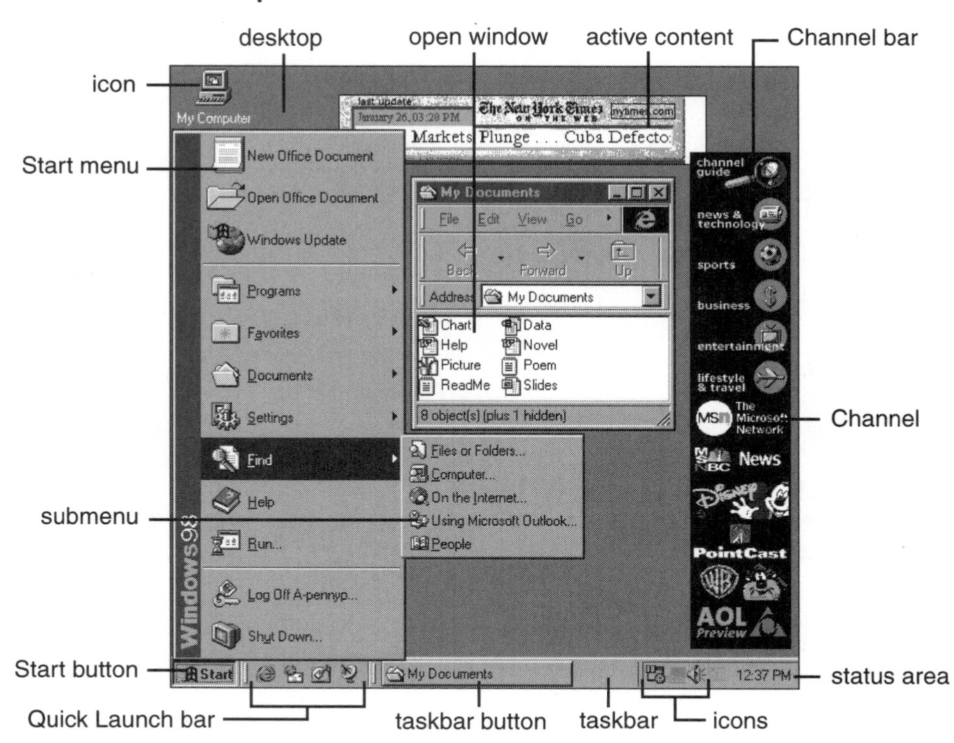

Open Window (Web view)

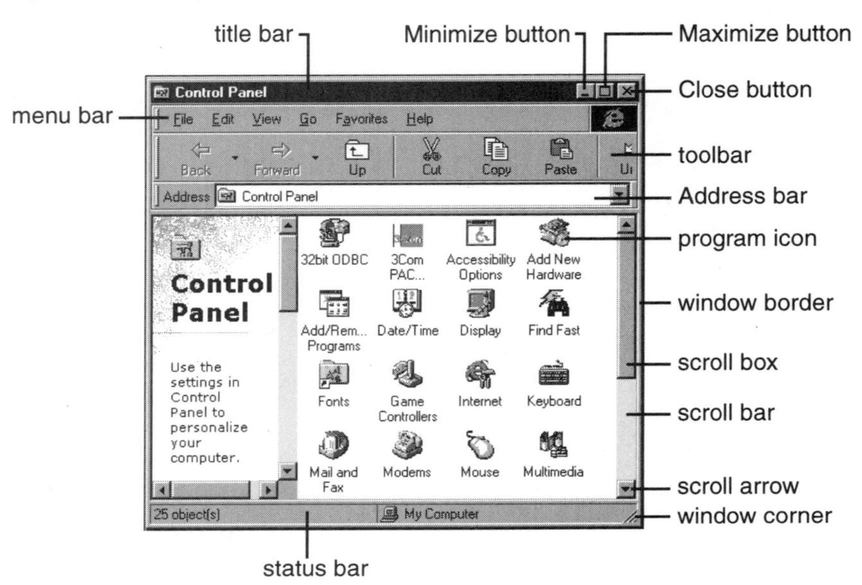

Browser

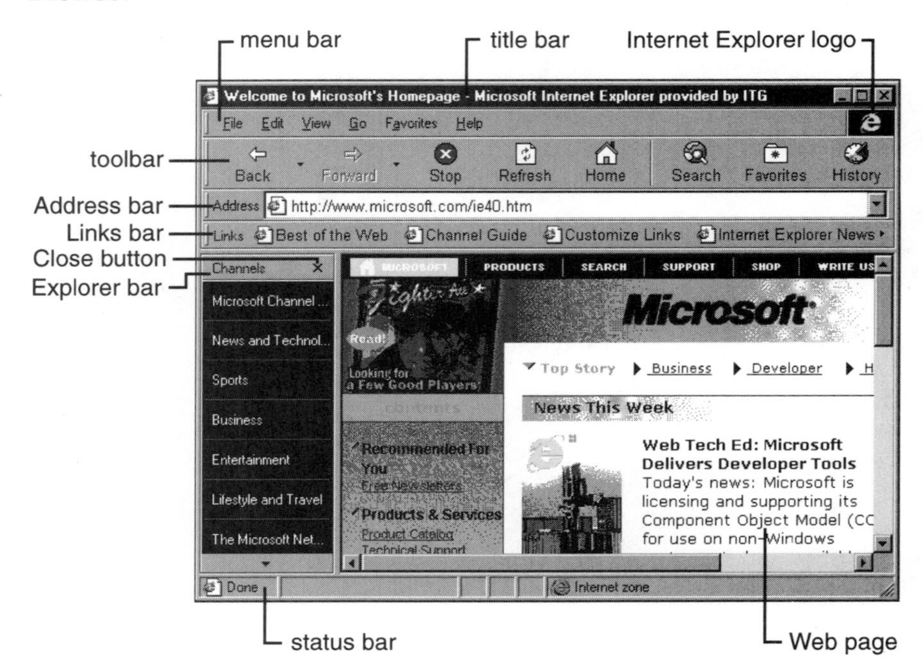

Web Page

Document Window

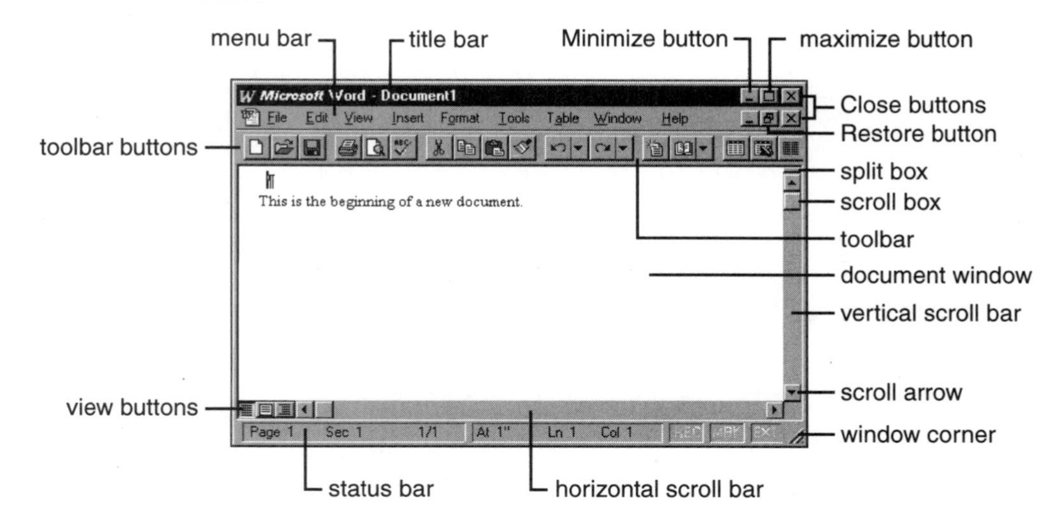

Menu with commands

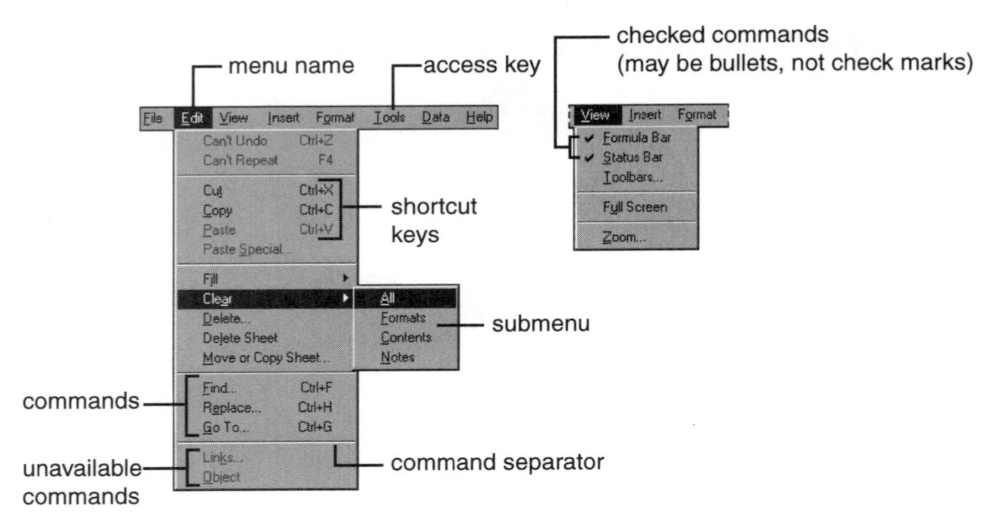

Macintosh desktop

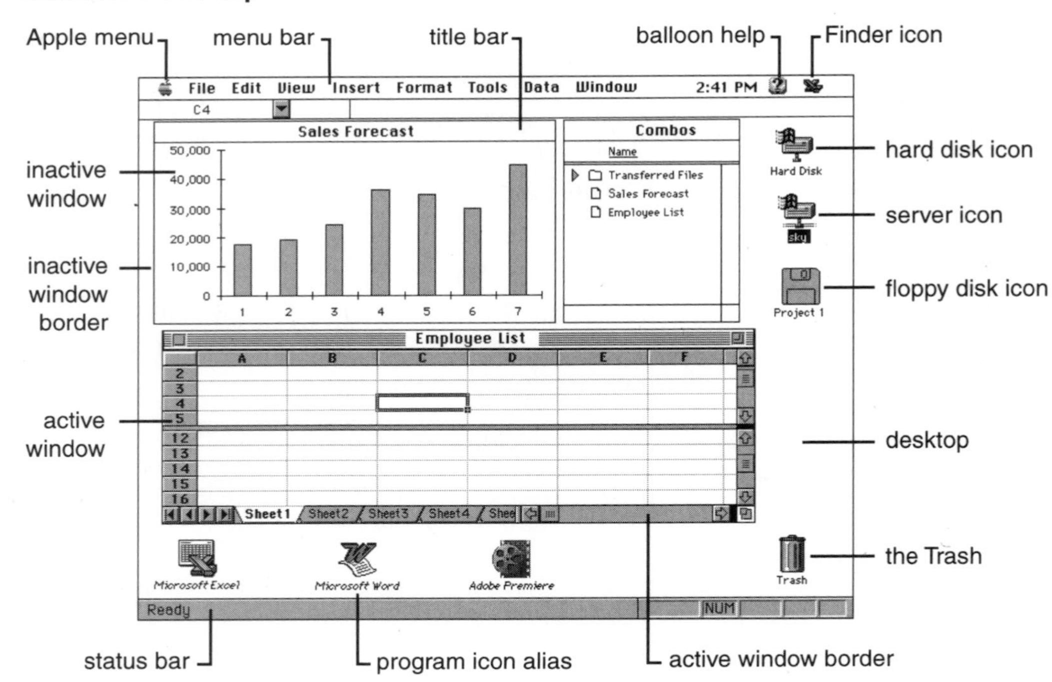

Macintosh document window

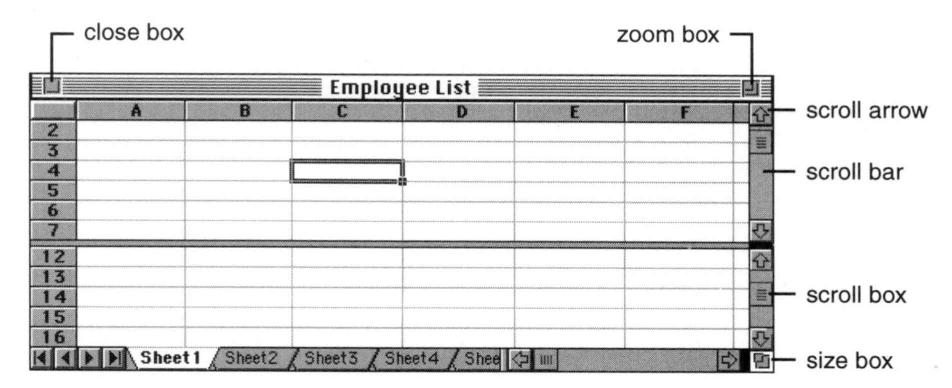

SEE ALSO **Dialog Boxes and Property Sheets, icon,** *The Windows Interface Guidelines for Software Design,* and the names of individual items

script, scripting language

Whenever possible, refer to a script generically. That is, just use *script* when you are referring to the code.

It is permissible to use *VBScript* occasionally to refer to the Visual Basic Scripting Edition, but the first reference to the product should refer specifically to the Visual Basic® Scripting Edition and there should also be subsequent references to it. Use phrasing such as "a script written in the Visual Basic Scripting Edition" or, later, "a script written in VBScript."

JScript, on the other hand, is a Microsoft trademark for development software. Do not refer to it as JavaScript (a Netscape product).

SEE ALSO **scriptlet**

scriptlet

Avoid in documentation unless you are specifically documenting "scriptlets." Just use *script,* defining it more specifically if necessary.

The word was probably formed by analogy with **applet**. Basically, a scriptlet is an ActiveX script control. It is programmed using dynamic HTML and runs in Internet Explorer 4.0. For more information, see //www.microsoft.com/mind/0198/cutting0198.htm.

scroll (v)

In general, treat *scroll* as part of a verb phrase rather than as a complete transitive verb. That is, use directional signals or prepositions with *scroll:* "scroll *through* the document," "scroll *left,*" "scroll *up* one line," "scroll *to* the next screen." If the concept of scrolling is already clear, use a verb phrase such as *move through.* Don't use *scroll* alone followed by a direct object, as in "scroll the window." However, it's acceptable to use it with a verb phrase such as "to see."

With the IntelliMouse pointing device, the user scrolls within the document by rotating the wheel of the mouse. It's acceptable to use directions such as "scroll up" in this instance.

Correct

You can scroll through the document to get to the end.
Drag the scroll box to move through the information.
Scroll until you see the new folder.

Incorrect

You can scroll the document to get to the end.
Drag the scroll box to scroll the information.

scroll arrow, scroll bar, scroll box

Use these terms specifically to refer to the interface element shown. Do not use *arrow* to refer to the *scroll arrow;* it can be confused with an up or down arrow.

Do not use *gray* or *shaded area* to refer to the *scroll bar.*

Do not use *slider* or *slider box* as a synonym for *scroll box.* See **slider**.

Scroll bar

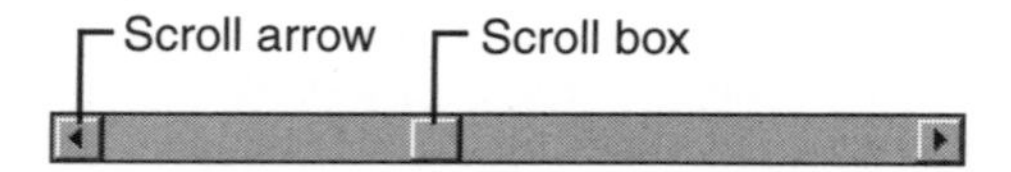

search, search and replace

Do not use for the search and substitution features; use *find* and *replace* instead.

Use *search, find,* and *replace* as verbs, not nouns. Avoid *search your document;* use *search through your document* instead.

Correct

Find the word "gem" and replace it with "jewel."
Search through your document for comments in red.

Incorrect

Do a search and replace.
Search your document for comments in red.

search engine

Okay to use in technical documentation to refer to a program that searches for keywords in documents or in a database. The term is now used most often in connection with services and programs that search the Internet using a particular search engine.

Avoid if possible in end-user documentation. Instead refer to the action itself ("Search for interesting Web sites") or use *search service* or *search page.*

Second Person

In general, use the second person *(you)* in most product documentation to refer to the user. Using the second person focuses the discussion on the user and makes it easier to avoid the passive voice. In material intended for developers, use the second person for the developer and use the third person *(the user)* for the developer's end user. Differentiate between *you* (the developer) and the program and actions it can perform.

Do not write in the first person *(I* or *we)* except in some specific circumstances (see **First Person**).

Correct

Just because you use Windows to get your work done doesn't mean you can't have fun too.
You can use Help topics to present information about your program in a format that users can access easily.
You can use this function in your program to allocate memory.
Your program can call this function to allocate memory.
Now Windows is faster and easier to set up.

Incorrect

Just because Windows is used to get work done doesn't mean it can't be used for fun too.
You can use Help to present information about your program in a format that can be accessed easily.
You can call this function to allocate memory.
We've made Windows faster and easier to set up.

Avoid writing error messages in the second person; by making the user the subject, it can sound as if the user is being blamed for the error. However, second person is useful in text that provides a solution or asks a question.

Correct

The printer on LPT1 is not responding.
Try restarting your computer.
Text outside the margins may not print. Do you want to continue anyway?

Incorrect

You are unable to print on LPT1.
The computer should now be restarted.

SEE ALSO **Active Voice vs. Passive Voice, Error Messages, First Person, should**

secondary menu

Avoid; use *submenu* instead. See **submenu**.

select, selected, selection

Use to refer to marking text, cells, and similar items that will be subject to a user action, such as copying text. Items so marked are called *the selection* or *selected.* Also, use to refer to dialog box options if you are documenting both keyboard and mouse use and to check boxes in all documentation. If you are documenting use of the mouse only, use *click* in these circumstances, except for check boxes.

In most programs, use *select,* not *highlight,* because selecting is a standard procedure, whereas highlighting could cause confusion in many programs. See **highlight**.

selection cursor

In general, avoid this term, which refers to the marker that shows where the user is working in a window or dialog box or shows what is selected. Use *insertion point* or *pointer* instead.

semicolons

Use semicolons sparingly, especially in end-user documentation—they seem formal and may indicate a convoluted sentence that should be rewritten. Unless the independent clauses are very closely related, it may be better to use two sentences instead. If necessary to save space online, you can use a semicolon in a procedure to replace a comma and coordinate conjunction. Be aware, however, that semicolons are difficult to see online.

Correct

Click All, and then press ENTER.

Correct to save space

Click All; then press ENTER.

Correct and appropriate for technical audience

The differences between assignment and initialization are:

- An assignment occurs when the value of an existing object is changed; an object can be assigned new values many times.
- An initialization occurs when an object is given an initial value when it is first declared; an object can be initialized only once.

It is correct to use semicolons within a sentence to separate phrases that contain other internal punctuation, especially commas. However, in documentation it may be clearer to the user if you break up such sentences into separate sentences or a bulleted list.

Convoluted

In this tutorial, you'll learn to quickly construct a user interface; easily implement both single-document interface and multiple-document interface applications; implement features that until now were considered difficult, such as printing, toolbars, scrolling, splitter windows, print preview, and context-sensitive help; and take advantage of many built-in components of the class library.

Better

In this tutorial, you'll learn to:

- Quickly construct a user interface.
- Easily implement both single-document interface and multiple-document interface applications.
- Implement features that until now were considered difficult, such as printing, toolbars, scrolling, splitter windows, print preview, and context-sensitive help.
- Take advantage of many built-in components of the class library.

Formatting semicolons

In printed documentation, set semicolons in the character formatting (roman, bold, or italic) of the preceding word unless the preceding word is a command, option, keyword, or placeholder; part of a code sample; or user input. In these cases, the semicolon takes roman formatting.

In online documentation, semicolons follow standard paragraph formatting following a jump or pop-up text.

setting

Use *setting* or *value* to refer to a specific value that the user can set, such as the specific color of a font ("you can choose blue as the setting for your font"). Differentiate from *attribute,* which is the general characteristic that can be set ("color is one attribute of fonts"). Avoid *property.*

set-top box

Standard industry term for the computer that sits on top of a television set to create two-way communication. Note hyphen. Do not use the abbreviation *STB.*

set up (v), setup (adj, n), Setup (the program)

Note the spelling according to grammatical function. Always spell the verb as two words; never hyphenate it.

Verb

Have everything unpacked before you set up your computer.

Adjective

The setup time should be about 15 minutes.

Noun

Your office setup should be ergonomically designed.
Run Setup before you open other programs.
Insert Setup Disk 1 in drive A.

SEE ALSO **install**

shaded

Use *shaded,* not *grayed* or *dimmed*, to describe the appearance of a check box that may occur when there is a mixture of settings for a selection in a group of options. The shaded appearance indicates that some previously checked options may make parts of the selection different from the rest.

Shaded options

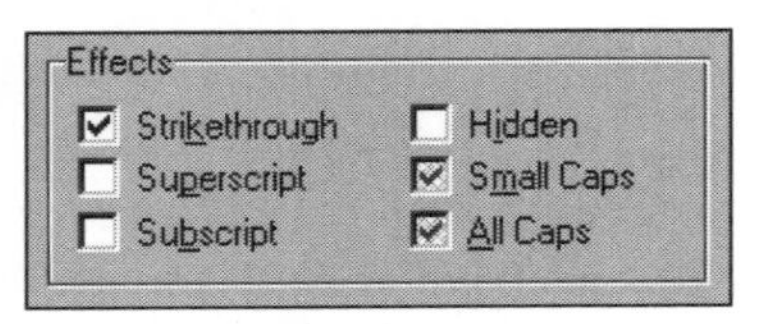

SEE ALSO **dimmed, gray, grayed**

shell (n)

Acceptable as a noun in technical documentation. Avoid in end-user documents. Avoid using *shell* or *shell out* as a verb—it's jargon. Use more precise terminology instead, such as "create a new shell" or "return to the operating system."

shortcut (adj, n)

One word, lowercase, to refer to shortcuts, shortcut buttons, and shortcut menus. Exception: Use initial cap for the Outlook Shortcuts that appear on the Office Shortcut Bar.

Do not use as a synonym for hyperlink.

shortcut key

Refers to a keyboard key or key combination such as CTRL+N or CTRL+S that invokes a particular command. Use this term in all documentation, including technical documentation in which you are referring to the interface. In discussions about programming the keys, you can use *accelerator key.*

If you need to specify the term, use the singular form when only one key is required: "Press the shortcut key F1 for Help." Otherwise, use the plural: "Press the shortcut keys CTRL+N to open a new file." But in general avoid the term, using the key name or key combination only: "Press F1 for Help."

SEE ALSO **Key Names**

shortcut menu

The shortcut menu appears when the user right-clicks a selection, a toolbar, or a taskbar button, for example. It lists commands pertaining only to that screen region or selection.

Shortcut menu

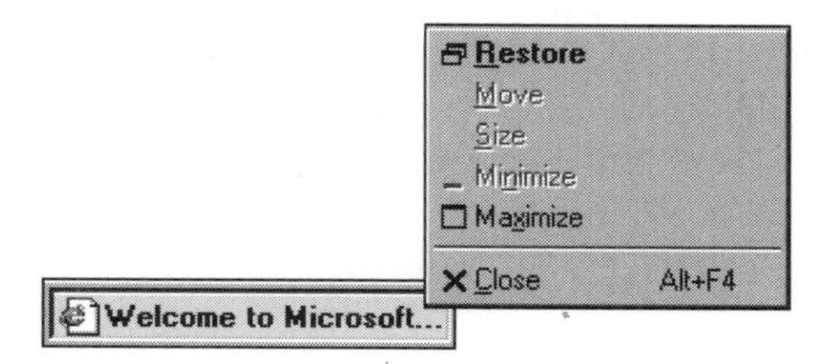

Use the term *shortcut menu,* not *context menu* or *pop-up menu*. In general, do not name the kind of shortcut menu, except for the toolbar menu.

Correct

... click Maximize on the shortcut menu.
... click Standard on the toolbar shortcut menu.

Incorrect

... click Maximize on the Word taskbar button menu.

should

Because *should* can be ambiguous, avoid its use. Instead, cast your sentence in one of the following ways, depending on the context:

- Use the imperative mood.
- Use *must* to specifically instruct users that they must follow some course of action.
- Use a phrase such as "we recommend" (in marketing information only) or "it is recommended."
- Rephrase the instruction to recommend some action or condition.

Correct, using imperative mood

Quit all programs before you shut down your computer.
Before you shut down your computer, save all your documents and quit all programs.

Correct, using must

You must register with Microsoft to receive free technical support.

Correct, stating a recommendation or condition

We recommend at least 8 MB of RAM to run this program.
If you want to save different copies of the document, you can save it under different file names.

Incorrect

You should quit all programs before you shut down your computer.
You should register with Microsoft to receive free technical support.
You should have at least 8 MB of RAM to run this program.
To save different copies of the document, you should save it under different file names.

A B C D E F G H I J K L M N O P Q R S T U V W X Y Z

shut down (v), shutdown (adj, n), Shut Down (command)

Do not use as a synonym for turning off the power to a computer; instead, use it to refer to the process of quitting all programs before turning off the computer.

Correct

Shut down your computer before you turn it off.

Note the spelling according to grammatical function. When referring to the command on the Windows Start menu, use two words and capitalize both (Shut Down).

Verb

Quit all programs before you shut down your computer.

Adjective

The shutdown procedure is simple.

Noun

The accidental shutdown corrupted some files.
Before you turn off your computer, point to Shut Down, and then click Shut down the computer?

sign in, sign off, sign on, sign up

Avoid *sign in* or *sign on* to refer to making a connection to the Internet. Use *make a connection* or a similar phrase instead. However, if you must use one, use *sign in,* not *sign on.* (Use *log on* for network connections.)

Avoid *sign off.* Use *disconnect* instead. Use *sign off* informally only to refer to getting approval.

Use *sign up* to refer to enrolling in a service.

Hyphenate these terms only when they are used as adjectives, not as verb phrases.

Correct

Type your sign-in name here.
Sign in here.
You can sign up for Internet Explorer by filling in the following information.
Will you sign off on this proposal?
We finally received the last sign-off we needed to move ahead on this project.

SEE ALSO **log on to, log off from, logon**

signed code

Technical term. Software that has obtained a digital certificate that the code has not been tampered with or will not be tampered with while it is being downloaded from the Internet. The certificate also identifies the author and software publisher, so the user can contact them if the code is not satisfactory.

simply

Avoid. It's generally unnecessary and can sound condescending.

site

Collection of Web pages developed as part of a whole, such as the Microsoft Web site, the Microsoft Library Web site, and so on. Preface with *Web* if necessary for clarity.

Information is "on" a Web site, but the address of a site is "at" http://www and so on.

SEE ALSO **page**

site map vs. sitemap

Use two words, by analogy with *American Heritage Dictionary*'s preferred spelling for *road map.*

size

Acceptable as a verb, as in "size the window." It is also acceptable to use *resize* to mean "change the size of."

size box

Use to refer to the small square in the lower-right corner of the Macintosh interface or MS-DOS-based programs only. Do not use to refer to the Maximize, Minimize, or Restore button in the Windows interface or to sizing handles.

SEE ALSO **Screen Terminology**

Slang

Avoid the use of slang in all documentation. The informal tone of slang is inappropriate in most documentation. Slang can also cause problems for localization. The line between slang and jargon (which can be acceptable in some circumstances) is often unclear, but slang is usually more informal and uses coinages and established words in new ways.

For information about specific terms, see the list of words under "Slang" in the index.

SEE ALSO **Jargon**

slash mark (/)

Do not use constructions containing a slash mark to indicate a choice, such as *either/or* and *he/she* (*and/or* is acceptable, if necessary). Never use a slash mark to separate two verbs, as in "open the File menu and open/close the file."

A B C D E F G H I J K L M N O P Q R S T U V W X Y Z

Use a slash mark in constructions that imply a combination.

Correct

client/server
CR/LF, carriage return/line feed
on/off switch
read/write

Use a slash mark to separate parts of an Internet address (use a double slash after the protocol name): *http://mslibrary/catalog/collect.htm.* Use a backslash with server, folder, and file names: *mslibrary**catalog**collect.doc.*

To refer to a slash in documents, it may be useful to differentiate between a "forward slash" for URLs and a "backward slash" for servers and folders.

You can also use a slash mark between the numerator and denominator of fractions in equations that occur in text. The Word Equation Editor includes a format with a slash mark.

Correct

$a/x + b/y = 1$
$x + 2/3(y) = m$

SEE ALSO **Numbers, Special Characters** (Appendix B)

slider

Control that lets users set a value on a continuous range of possible values, such as screen brightness, mouse-click speed, or volume. Do not use to refer to the scroll box or a progress indicator.

Slider

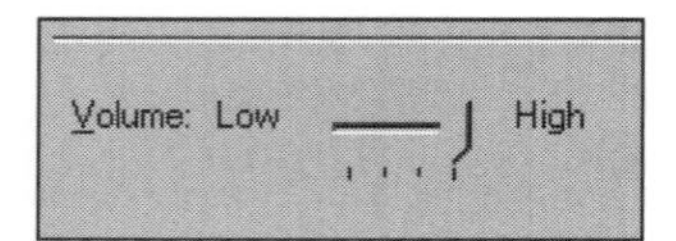

Correct

Move the slider to the left to decrease pointer speed.

SEE ALSO **Dialog Boxes and Property Sheets, progress indicator**

small caps

Do not use small caps in documentation for key names or A.M. and P.M. They are awkward to designate in code such as HTML.

You can use the term *small caps* in documents. If necessary for clarity, refer to them as *small capitals* first, followed by a phrase such as "often referred to as 'small caps.' "

smart cards, Smart Cards

Physically, a smart card is a card with an embedded integrated circuit for storing information. Smart cards are a key component of the public-key infrastructure Microsoft is integrating into the Windows and Windows NT systems.

Use lowercase for generic references to the cards or technology. Use initial caps to refer to the Microsoft product and architecture. For more information, see //www.microsoft.com/smartcard/.

soft copy

Avoid; it's jargon formed by analogy with *hard copy.* Instead, use a more specific term such as *electronic document* or *file.*

spam

Internet jargon referring to littering large numbers of e-mail boxes with unwanted messages. Avoid in most documentation; use a term such as "junk e-mail" instead. Do, however, define it and include it as a keyword if appropriate.

specification

Capitalize *specification* when it is part of the title of a document, as in *The Network Driver Interface Specification.* Avoid the informal *spec* in documentation.

speed key

Do not use. Use *access key* or *shortcut key* instead, depending on meaning.

SEE ALSO **access key, Key Names, shortcut key**

spelling checker

Refer to the tool as the *spelling checker,* not *spell checker* or *Spell Checker.* Do not use *spell check* as a verb or noun.

Correct

Use the spelling checker to check the spelling in the document.

Incorrect

Spell check the document.
Run the spell checker.

spider

Refers to an automated program that searches the Internet for new Web documents and places information about them in a database that can be accessed by a search engine. Okay to use, but define first.

spin box

Control that lets users move ("spin") through a fixed set of values, such as dates. Use only in technical documentation. Use the name of the specific box in end-user documentation—for example, the Start time box.

Spin box

SEE ALSO **Dialog Boxes and Property Sheets**

split bar

Refers to the horizontal or vertical double line that separates a window into two panes. In some programs, such as Windows Explorer, the window is already split, but the user can change the size of the panes. In other programs, such as Word or Microsoft Excel, the user can split the window by using the split box. The term is acceptable in both technical and end-user documentation.

Split bar separating two panes

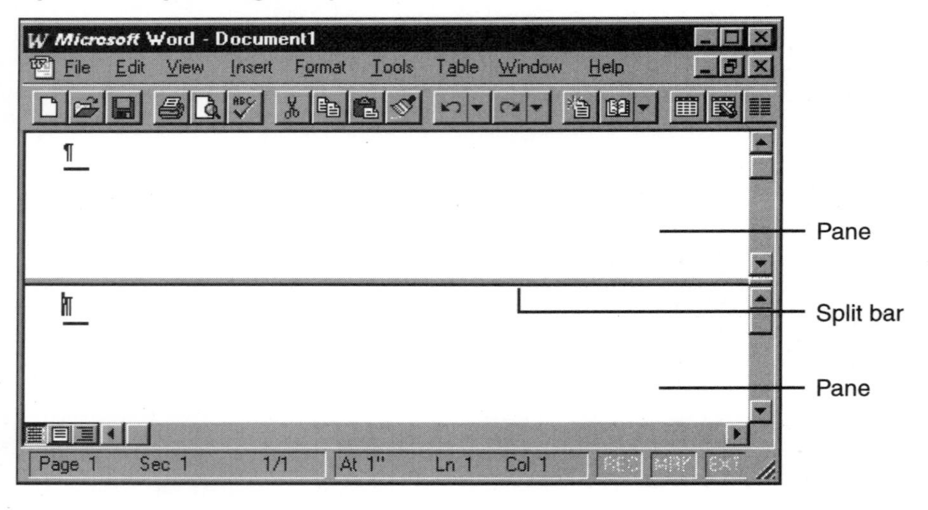

SEE ALSO **Screen Terminology**

split box

Refers to the control at the top right of the vertical scroll bar (for horizontal splitting) or far right of the horizontal scroll bar (for vertical splitting). Users point to the split box to drag the split bar. The term is acceptable in all documentation.

Split box on a horizontal scroll bar

spoofing

Refers to the practice of tricking users into providing passwords and other information to allow unauthorized access into a system. Although it is a commonly used term, avoid in most documentation. If it's appropriate to the product, define it at first use.

spreadsheet

Do not use as a synonym for *worksheet.* A spreadsheet is a computer accounting program, such as Microsoft Excel; a worksheet is the document produced by a spreadsheet program.

SQL Server

SQL Server is the name of the Microsoft product. At first mention and occasionally thereafter within a document, use *Microsoft SQL Server*. When referring to a computer running Microsoft SQL Server, use *the SQL Server* or *SQL Servers* (note the capital *S*).

When referring to the product or the server running the product, *SQL* is pronounced "sequel" and takes the article *a* when not preceded by *Microsoft*—for example, "a SQL Server." When referring to the language SQL (SQL stands for Structured Query Language), SQL is pronounced "es-cue-el" and takes the article *an*—for example, "an SQL database."

> **NOTE** It's acceptable to use the redundant term "SQL language" if necessary.

stand-alone (adj)

Do not use as a noun.

Correct

You can use Word either as a stand-alone word processor or on a network.

Incorrect

Some early word processors were stand-alones.

start, Start (the menu)

In general, use *start,* as in "start a program," instead of **boot**, **initiate**, **initialize**, **issue**, **launch**, **turn on** , and so on. Capitalize references to the Start menu and the Start button on the taskbar, and always specify which one you are referring to.

Avoid "click the Windows Start button."

Correct

Start Windows, and then click the Start button to start your programs.
On the taskbar, click the Start button, and then click Run.

start page

Refers to a Web page designated by the user as the first page seen when opening a browser. Do not use. It has been replaced by *home page* in Microsoft Internet Explorer. See **home page**.

A B C D E F G H I J K L M N O P Q R S T U V W X Y Z

status bar

Not *status line* or *message area.* Refers to the area at the bottom of a document window that lists the status of a document and gives other information, such as the meaning of a command. Messages appear on, not in, the status bar.

Correct

The page number is displayed on the status bar.

Writing status messages

Follow these guidelines for writing effective status bar messages.

- Use parallel constructions and begin the message with a verb.

  The message describing the View menu, for example, should read something like "Contains commands for customizing this window" and the message describing the Internet folder icon should read something like "Changes your Internet settings."
- Use present tense.

  For example, use "Changes your Internet settings," not "Change your Internet settings."
- Make sure the text is constructive. Try to avoid repeating the obvious.

  For example, even though the File menu is quite basic, a message such as "Contains commands for working with the selected items" gives some useful information with the inclusion of the phrase "selected items."
- Use complete sentences, including articles, and end with a period.

SEE ALSO **Screen Terminology**

stop

Acceptable to use to refer to hardware operations, as in "stop a print job." Use *quit* with programs.

storage, storage device

Do not use *storage* to refer to memory capability; use *disk space* instead. *Storage device* is acceptable as a generic term to refer to things such as disk and tape drives.

stream, streaming

Okay to use either term as a noun or verb, depending on context and meaning, to refer to video or other graphics coming to a browser over the Internet. A stream is also an I/O management term in C.

Avoid in other metaphoric uses if possible.

strike

Do not use to refer to keyboard input; use *press* or *type* instead.

SEE ALSO **press**

A B C D E F G H I J K L M N O P Q R S T U V W X Y Z

strikethrough

Not *strikeout* or *lineout.* Refers to the line crossing out words in revisions.

struct

Do not use in normal text to refer to a data structure in code. It's jargon. Use *structure* instead.

style sheet

Two words. Can refer to a file of instructions for formatting a document in word processing or desktop publishing or to a list of words and phrases and how they are used or spelled in a particular document.

In Internet use, refers to a cascading style sheet (a .css file) attached to an HTML document that controls the formatting of tags on Web pages. The browser follows rules (a "cascading order") to determine precedence and resolve conflicts.

For more information about cascading style sheets, see //www.microsoft.com/workshop/author/css/css-f.htm.

sub (prefix)

In general, do not hyphenate words beginning with *sub,* such as *subheading* and *subsection,* unless it's necessary to avoid confusion or *sub* is followed by a proper noun, as in *sub-Saharan.* If in doubt, check *American Heritage Dictionary* or your project style sheet.

subaddress

Do not use to refer to parts of an address that go to a specific place in a file, such as a bookmark. Use the specific term instead.

subclass

Do not use as a verb. It's jargon. Use a standard verb, such as *subordinate,* or a phrase, such as "make into a subclass."

Subjunctive Mood

There is seldom any reason to use the subjunctive mood, which expresses a condition contrary to fact or a wish, desire, supposition, or hypothesis. In general, write documentation in the indicative mood (for information) or the imperative mood (for procedures).

Unnecessary subjunctive

It's essential that the Shut Down program be run before the computer is turned off.

Indicative (better)

It's essential to run the Shut Down program before you turn off your computer.

Imperative (better)

Run the Shut Down program, and then turn off your computer.

SEE ALSO **Mood of Verbs**

submenu

Describes the secondary menu that appears when the user selects a command that includes a small arrow on the right. Avoid in end-user documentation if possible, for example, by referring only to what appears on the screen. The term is acceptable in programmer documentation.

Submenus

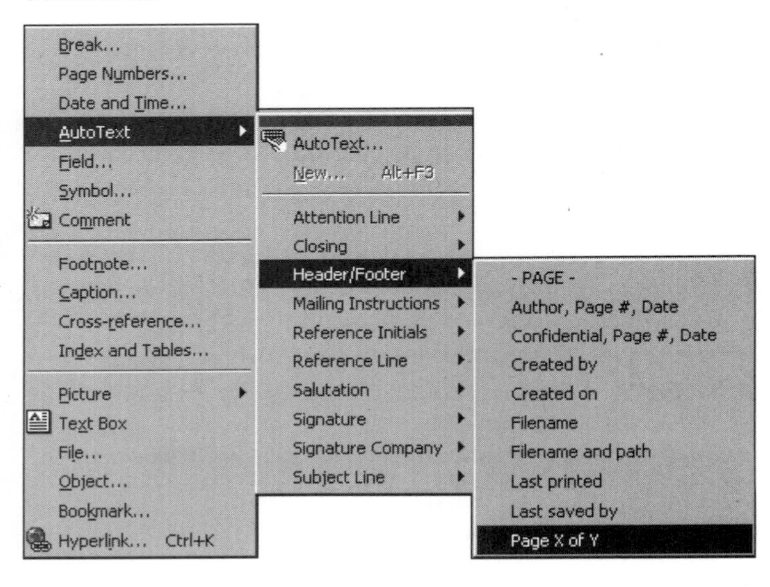

Avoid the terms *cascading menu, hierarchical menu,* and *secondary menu.*

Correct (in a procedure)

- On the Edit menu, point to Clear, and then click the item you want to clear.

SEE ALSO **Screen Terminology**

Super VGA, SVGA

Use *Super VGA (SVGA)* the first time you refer to a video adapter with higher resolution than a VGA. After that it's okay to use *SVGA.*

surf

To browse the Web. Generally implies a more random browsing than the less informal *browse.* Okay to use if accurate in informal documents.

switch

Acceptable to use as a verb, as in "switch to another window." Use instead of *activate* or *toggle.*

Correct

To embed the new object, switch to the source document.
You can easily switch between open windows.

Incorrect

To embed the new object, activate the source document.
You can easily toggle between open windows.

Acceptable as a noun in programming documentation to refer to command-line and compiler options, such as /Za. Because some groups use *option* instead, refer to your project style sheet.

symbol

Use the word *symbol* to refer to a graphic or special character that represents something else, but differentiate a symbol from an icon. (An icon represents an object the user can select and open. A symbol can appear on an icon.)

Follow these guidelines for discussing symbols:

- Write out the name of the symbol in text and, if the symbol itself is important, enclose the symbol in parentheses. Use symbols alone only in tables and lists where space is at a premium or in mathematical expressions.

  Correct
  You can type a backslash (\) to return to the previous directory.
  Only 75 percent of the students attended.

  Incorrect
  You can type a \ to return to the previous directory.
  Only 75% of the students attended.

- For screen elements such as buttons, you can use only a graphic of the button after it has been named once or if clicking it brings up a definition.

  Correct
  Click Minimize ( ).
  Click .

- Spell out *plus sign, minus sign, hyphen, period,* and *comma* when referring to them as key names.

  Correct
  Press COMMA.
  Type a comma.
  Press the PLUS SIGN (+).

  Incorrect
  Press ,.
  Press +.

- Write out plurals of symbols, showing the use in parentheses. Do not add *s* or *'s* to a symbol.

  Correct
  Type two backslashes (\\) to show a network connection.

  Incorrect
  Type two \'s to show a network connection.
  Type \\s to show a network connection.

- Do not insert a space between a number and the symbol it modifies.

  Correct (to conserve space)
  75%
  <100

  Incorrect
  75 %
  < 100

SEE ALSO **Measurements, Special Characters** (Appendix B)

sysop

Even though *sysop* is jargon for "system operator," it's acceptable to use to refer to the person who oversees or runs a bulletin board system or online communications system in documentation about such products. Define first. Avoid in all other documentation.

system

Use generically to refer to computer hardware configurations, not the computer alone. The system includes the computer and peripheral devices. It's not synonymous with, but can include, the *system software.*

system administrator

Use only to refer to the person responsible for administering the use of a multiuser computer system. Generally, use *administrator* unless you must specify a particular kind. Use *network administrator* only to specifically refer to the administrator of networks.

system prompt

Do not use; use *command prompt* instead. If necessary, be specific in naming the command, as in "MS-DOS prompt."

System Requirements

Minimum hardware and operating system requirements needed to run Microsoft software appear as a bulleted list or in a table in the introduction to the main printed book or Getting Started topic and on the product packaging. (Creative Services handles the list on the packaging.) For operating systems and programming products, you can include hardware architecture and examples of vendor hardware if necessary.

Use the guidelines in the following sections for the kinds of equipment to include and the order of the list. You can combine the operating system and processor requirements to shorten or clarify the list.

Use trademark bugs as legally required if the list contains the first mention of a trademarked product. See **Trademarks**.

For Windows-based and MS-DOS-based software, do not include references to personal computer vendors (for example, IBM personal computers). For Macintosh products, however, do include Apple-specific references, which also cover clones.

Include the following items for all types of software:

- Operating system, including both minimum and recommended system
- Processor required and recommended
- RAM required and recommended
- Disk drives
- Hard disk space required and recommended
- Other requirements and recommendations, such as a description of the monitor, video adapter, or CD-ROM drive, if applicable

Correct (list method)

Microsoft Internet Information Server requires the following minimum system configuration:

- Microsoft Windows NT operating system version 3.5 or later (Windows NT Server 4.0 or later recommended)
- Personal computer with a 486/66 or higher processor (Pentium 90 recommended) or a RISC-based system such as the MIPS R4000 or R4400
- 32 megabytes (MB) of RAM (64 MB recommended)
- 50 MB of free hard disk space (minimum installation; 200 MB recommended)
- VGA monitor (Super VGA recommended)
- Optional 3X CD-ROM drive (6X recommended)

Correct (table method)

Microsoft Internet Information Server requires the following system configuration:

Component	*Requirement*	*Recommendation*
Operating system	Windows NT 3.5	Windows NT 4.0
Processor	66 MHz 486	90 MHz Pentium
Free hard disk space	32 MB	64 MB
Monitor	VGA	Super VGA
CD-ROM drive (optional)	3X	6X

system software

Not *systems software.* Use generically to refer to an operating system or software that extends operating system functionality.

tab (n, adj)

Do not use as a verb. Because multiple uses can be ambiguous, especially for localization, use the noun *tab* alone to refer only to a tab in a dialog box. For other uses, clarify the meaning with a descriptor: the TAB key, a tab stop, or a tab mark on the ruler.

Correct

Use the TAB key to move through a dialog box.
Set a tab stop on the ruler.
Click the View tab.

Incorrect

You can tab through a dialog box.
Set a tab on the ruler.

table of contents

Do not use *Table of Contents* as the heading for the list of contents at the beginning of a document or file; use just *Contents* instead. It's correct to refer generically to the *table of contents,* however.

Tables

A table is an arrangement of data with two or more columns in which the information in the first column relates specifically to the information in the other column or columns. (A list of similar entries that is arranged in multiple columns is not a table but a multicolumn list — for example, a list of commands.) The format of a table can vary, depending on the project style.

A table usually has column headings and may or may not have a title. Introduce tables with a sentence ending with a period. If a table is titled, an introductory sentence does not have to immediately precede the table.

Tables on the Web should follow the same guidelines as tables in other documentation. Keep in mind these additional points, however:

- Table dimensions must be planned to not exceed the viewing area of various screen resolutions.
- Tables can cause screen readers to form text into columns, ignoring the table column format. That in turn would be confusing to a blind person.
- Tables can be used to simulate frames. In this case, tables are better because more browsers can correctly interpret tables than they can frames.

A B C D E F G H I J K L M N O P Q R S T U V W X Y Z

Capitalization and punctuation

If the table is titled, use title caps for the title. That is, do not capitalize articles *(a, an, the),* prepositions of four or fewer letters, or coordinate conjunctions. Capitalize only the first word of each column heading, the first word of each column entry, and proper nouns.

NOTE You can use lowercase for the first word in column entries if capitalization might cause confusion — for example, a column of keywords that must be lowercase.

End each entry with a period if all entries are complete sentences or are a mixture of fragments and sentences. An exception is when all entries are short imperative sentences (only a few words); these entries do not need a period. If all entries are fragments, do not end them with periods.

Number tables in printed documents if they are not adjacent to the text reference or if a list of tables appears in the front matter. Do not number online tables. If you decide to number tables, use numbers consistently throughout the document. The numbers include the chapter number and a sequential table number, such as Table 2.1, Table 2.2, and so on.

Correct

Table 7.4 Formatting Flags

Formatting flag	*Action*
\a	Anchors text for cross-references
\b, \B	Turns bold on or off

Content

Tables read across from left to right. That is, information in the right column relates horizontally to the information in the left columns. Column headings should reflect this flow.

- Place information that is known or obvious to the user in the leftmost column of the table. Place information that the user is to learn or an action the user is to perform in the subsequent columns.
- Make entries in a table parallel — for example, in a description column, be consistent in your use of beginning the entries with a verb or noun.

Correct

Device name	*Description*
COM1	Serial port 1. This device is the first serial port in your computer.
CON	System console. This device consists of both the keyboard and the screen.
LPT1	Parallel port 1. This device represents a parallel port.

Command	*Action*
Bold	Turns bold on or off
Italic	Turns italic on or off

To move the insertion point	*Press*
To the first record	The TAB key
To a record you specify	ENTER

Incorrect

To	*Do This*
Close a window	Use the Minimize button.
Size a window	Press CTRL+F8
Copy a picture	The TAB key

- Do not leave a blank column entry. That is, if the information doesn't apply, use *Not applicable* or *None.* Do not use em dashes.

Correct

To	*In Windows, press*	*On the Macintosh, press*
Copy a picture	None	COMMAND+SHIFT+T

Column headings

Make column headings as concise and as precise as possible, but include information common to all column entries in the heading, rather than repeat it in each entry. Do not use ellipses.

Correct

To	*Do this*
Open a Web page	Type the address in the Address bar, and then press ENTER.
Add a Web page to your favorites list	Click Favorites, and then click Add to Favorites.

To save a document	*Do this*
To a folder	Click Save.
Under a new name	Click Save As.
To a network location	Connect to the server location and folder, and then click Save.

Voice

In tables that list procedures, use the active voice in column headings, preferably in phrases that reduce repetition in the entries in the table.

Correct

To	*Do this*
Close a window	Click the Minimize button.
Size a window	Press CTRL+F8.

Formatting

Most table formatting is done within your design template. These guidelines suggest ways to make tables more readable.

- In printed documents, try to limit tables with long entries to two or three columns. Four or more columns can be hard to format and read unless they contain brief, numeric entries. The second column in the following example is approaching maximum readable length.

Table 2.2 Addressing Declared with Microsoft Keywords

Keyword	*Data*	*Code*	*Arithmetic*
_ _near	Data resides in the default data segment; addresses are 16 bits.	Functions reside in the current code segment; addresses are 16 bits.	16 bits
_ _far	Data can be anywhere in memory, not necessarily in the default data segment; addresses are 32 bits.	Functions can be called from anywhere in memory; addresses are 32 bits.	16 bits
_ _ huge	Data can be anywhere in memory, not necessarily in the default data segment. Individual data items (arrays) can exceed 64K in size; addresses are 32 bits.	Not applicable; code cannot be declared _ _huge.	32 bits (data only)

- Try to avoid dividing a table between pages. If a titled table is continued on another page, repeat its title followed by *(continued)* (all lowercase, in parentheses, and italic, including the parentheses) and repeat its column headings. For an untitled table, repeat its column headings. If a table is very long, consider breaking the material into logically related subtables.
- Use rules between rows if the column information varies.

Footnotes

Your choice of footnote designator depends on the material in the table. For example, if the table contains statistical information, use a letter or symbol to designate a footnote. The order of preference for footnote symbols is numbers, letters, and symbols. For a list of symbols, see *The Chicago Manual of Style.*

Put footnote explanations at the end of the table, not the bottom of the page.

SEE ALSO **Lists**

Taiwan

Use *Taiwan* or *Taiwan Region,* not *Republic of China* or *ROC,* to refer to this province of the country of China.

tap

Use *tap* (and *double-tap*) instead of *click* when documenting procedures specific to pen pointing devices. *Tap* means to press the screen and then lift the pen tip.

target disk, target drive, target file

Do not use; it can be ambiguous and difficult to translate accurately. Use *destination disk, destination drive,* or *destination file* instead.

taskbar

The bar that appears by default at the bottom of the Windows 95 and later desktop.

Taskbar

Users click buttons that appear on the taskbar to switch between running programs. See **Screen Terminology**.

TB

Do not use as an abbreviation for *terabyte,* which should not be abbreviated. See **terabyte**.

Telephone Numbers

Because telephone numbers change frequently, this guide does not list specific phone numbers and when to use them. Instead, it contains guidelines for handling and punctuation.

General guidelines

Use parentheses, not a hyphen, to separate the area code from the seven-digit phone number. Do not precede the area code with 1 except in international lists.

Correct

(206) 882-8080
(800) 555-0123

Incorrect

206-882-8080
1-800-555-0123

Some 800 (toll-free) numbers are accessible to both U.S. and Canadian callers, and some serve only either U.S. or Canadian callers. If a number serves only one country, be sure to indicate that the number or the service it provides is not available outside that country.

International phone numbers

In international phone numbers, use parentheses, not hyphens or spaces, to separate the country and city codes (one- to three-digit numbers that are equivalent to the U.S. area code) from the local

A B C D E F G H I J K L M N O P Q R S T U V W X Y Z

phone number. Use hyphens to separate the parts of the local phone number, not spaces or parentheses. In the following example, (44) is the country code for the United Kingdom, (71) is the city code for London, and 270-0012 is the local phone number.

Correct

(44) (71) 270-0012

Incorrect

(44) (71) 270 0012
44-71-270-0012

You can find a list of international area codes in your local telephone directory.

In lists of several international numbers, include country and city codes for any number that might be dialed from outside that country. Many numbers serve callers from outside that country; for instance, many African countries are served by the subsidiary in South Africa, so these callers need the South African country and city codes. This means that you will virtually always provide the country and city codes.

When U.S. numbers are included in lists of international numbers, include (1), which is the U.S. country code. There are two reasons for this: It provides the U.S. country code to callers outside the United States, Canada, and the Caribbean, and it keeps the format consistent for all numbers in a list.

In international numbers, do not include a country's long-distance access code, which is the number callers inside that country use to initiate an international long-distance call. (For example, the number 011 is the AT&T access code within the United States.) Do not put a plus sign (+), indicating the need for an access code, in front of the number.

Correct

(81) (3) 513-888
(206) 882-8080 [when only domestic numbers are provided]
(1) (206) 882-8080 [when both domestic and international numbers are provided]

Incorrect

+(81) (3) 513-888
011-81-3-513-888
(1) (206) 882-8080 [when only domestic numbers are provided]

Fictitious phone numbers

For fictitious phone numbers, use the prefix 555 and four digits between 0100-0199— for example, 555-0187. These numbers are not assigned to any lines. Fictitious numbers are covered in the boilerplate copyright notice, so no special mention on the copyright page is required.

telnet, Telnet

Use lowercase to refer to a client program that implements the Telnet terminal-emulation protocol or to using the protocol. Use initial cap to refer to the protocol itself. In UNIX usage, the protocol is usually all caps (TELNET).

tense

Use simple present tense. Try to avoid all other tenses. Present tense helps readers scan the material quickly.

Correct

Although the Microsoft Mail system is very reliable, it's a good idea to back up important messages periodically.
The next section describes how to write an object-oriented program.

Incorrect

Although the Microsoft Mail system has proven to be very reliable, it's a good idea to back up important messages periodically.
The next section will describe how to write an object-oriented program.

Use past and future tenses only when it is confusing to use the present tense — for example, when it's essential to describe events in terms of the past or future, as when a future event will be caused by a present action.

Correct

If you are going to use the macro as a demonstration, you will want that macro to play back at the same speed at which it was recorded.

terabyte

One terabyte is equal to 1,099,511,627,776 bytes, or 1,024 gigabytes.

- Do not abbreviate.
- Leave a space between the numeral and *terabyte* except when the measurement is used as an adjective preceding a noun. In that case, use a hyphen.

Correct

36 terabytes
36-terabyte database

- When used as a noun in measurements, add *of* to form a prepositional phrase.

Correct

This database contains 36 terabytes of information.

SEE ALSO **Measurements**

terminal

Do not use. Terminals, which generally do no processing on their own, are seldom used in personal computer systems, and the term can be ambiguous. Use *workstation* or *computer* instead. The term *terminal-emulation program* is acceptable.

SEE ALSO **workstation**

terminate

Do not use; use *quit* or *close* instead.

SEE ALSO **close, quit**

text box

When referring to dialog boxes, use *text box* to refer to the area where the user can type text. Do not use *entry field,* which is an acceptable term only in database programs. See **Dialog Boxes and Property Sheets**.

Text box

that vs. which

That and *which* are often confused. Be sure to use the appropriate word. *That* introduces a restrictive clause, which is a clause that is essential for the sentence to make sense. A restrictive clause often defines the noun or phrase preceding it and is not separated from it by a comma. In general, do include the word *that* in restrictive clauses, even though in some clauses the sentence may be clear without it. Including *that* prevents ambiguity and helps translators understand the sentence.

Correct

You will need to supply information about applications that you want to run with Windows.

Incorrect

You will need to supply information about applications which you want to run with Windows.
You will need to supply information about applications you want to run with Windows.

Which introduces a nonrestrictive clause, which is a clause that could be omitted without affecting the meaning of the sentence. It is preceded by a comma. Nonrestrictive clauses often contain auxiliary or parenthetical information.

Correct

Your package contains the subsidiary information card, which you can use to obtain device drivers or local technical support.

Do not use *that* or *which* to refer to a person; instead use *who.* See **who vs. that**.

then

Then is not a coordinate conjunction and thus cannot correctly join two independent clauses. Use *and* or another coordinate conjunction or *then* with a semicolon or another conjunctive adverb to connect independent clauses in, for example, two-part procedural steps.

Correct

On the File menu, click Save As, and then type the name of the file.

Incorrect

On the File menu, click Save As, then type the name of the file.

Avoid using *then* to introduce a subordinate clause that follows an *if* clause (an "if ...then" construction).

Correct

If you turn off the computer before shutting down all programs, you may lose data.

Incorrect

If you turn off the computer before shutting down all programs, then you may lose data.

thread (n)

Okay to use to refer to a series of articles or messages on the same topic in a newsgroup or e-mail discussion.

three-dimensional, 3-D (adj)

Three-dimensional is preferred, but *3-D* is acceptable. Spell out at first mention. Use *3-D* in tables and indexes and where space is a problem, as well as to reflect the interface.

Hyphenate both the spelled out and abbreviated versions. Use *3D* (no hyphen) only as specified by product names.

35mm

Abbreviation for 35-millimeter. Note that 35mm is one word, no hyphen. It is not necessary to spell out at first mention.

time-out (adj, n), time out (v)

Always hyphenate as an adjective or noun. Do not hyphenate as a verbal phrase.

Correct

A time-out occurs if the connection can't be made.
If the connection isn't made, a time-out event occurs.
The connection timed out.

Time Zones

Designations of time zones should be lowercase unless they contain a proper adjective. Do not specify daylight or standard time, because this status changes every six months, so any reference will quickly become obsolete. Do not use abbreviations, such as *PST, CDT,* or *EST,* for time zones.

Correct

central time
eastern time
Greenwich mean time
Pacific time

Incorrect

central daylight time
Eastern time
eastern standard time
GMT
pacific time

Tips

A tip helps users apply the techniques and procedures described in the text to their specific needs.

For more information about the use and purpose of tips, see **Notes and Tips**.

title bar

The horizontal bar at the top of a window that shows the name of the document or program. You can use this term in both end-user and technical documentation.

Title bar

titled vs. entitled

If you need to specify the title of a book, program, or other item, use the word *titled* instead of *entitled.* The word *titled* is not followed by a comma — for example, "Look in the manual titled *User's Guide,* which accompanies your software."

In general, however, avoid using *titled* as well. Instead, say, for example, "look in the *User's Guide* that accompanies your software."

Do not use *entitled* as a synonym for *titled;* instead, use to mean "is owed." Books are *titled;* a user may be *entitled* to a set of documentation.

Titles of Books

The title page of a printed book includes the product name and generally the product descriptor — for example, the title page for the Windows 95 user's guide includes this information:

Introducing Microsoft Windows 95
For the Microsoft Windows 95 Operating System

In the guide itself, the title is referred to as *Introducing Microsoft Windows 95.*

Thus, as long as the product name is prominently mentioned, it doesn't necessarily have to be included as part of the actual book title. For example, a book can be titled *Installation Guide* as long as the product name also appears. A "complete" title could be, for example, *Installation Guide* For Microsoft Exchange Server, if the product name appears on both the cover and title page.

NOTE Refer to a book as a *book,* not as a *manual.*

While this topic refers to "book titles," implying that all books are printed, most books now appear online in some form, as well. Online documentation can include any of these books.

The following table lists some common Microsoft titles, the audience for that type of book, and typical content. Users expect to find certain material in certain types of documentation regardless of what the product is. To maintain consistency across product lines, it is important to use the titles precisely and appropriately. The list is not complete — individual projects may need additional, unique titles. Do not, however, create a new title when one from the list will serve your purposes.

Titles of Books

Title	*Audience*	*Content*
Administrator's Guide	Technical support personnel, system and network administrators	Task-oriented information about configuring, installing, and managing a product.
Administrator's Reference	Technical support personnel, system and network administrators	Comprehensive, often encyclopedic information about the product features.
Companion	End users	Overview of product features, often describing projects the user can accomplish with the product, such as publishing a newsletter. Often highly visual and informal.
Conversion Guide	Programmers, administrators	Explanation of how to convert files or programs from one system to another. Not the same as a book that teaches users of one product how to use a similar Microsoft product.
Design Guide	Programmers, application developers, interface designers	Technical information about designing a program interface.
Developer's Guide	Technical users who may not be programmers, such as database and macro developers	Explanation of development concepts and techniques.
Feature Guide	End users	Overview of product features.

(continued)

Title	Audience	Content
Getting Results with [Microsoft Product Name]	End users	Task-oriented introduction to the product, possibly including installation. Focuses on helping users accomplish specific kinds of tasks.
Getting Started	End users, often novices	Basic installation and setup information. May include a road map to the product or other documentation, a summary of new features, and tutorial material.
Idea Book	End users	Task-oriented. Highlights certain features of the product. Often includes sample files.
Installation Guide	All users	Information about how to install the product.
Language Reference	Programmers	The complete syntax of a programming language. Usually includes extensive examples and usage notes.
Library Reference	Programmers	The functions that Microsoft ships with a programming language. Includes extensive examples, and may include sample files. Often alphabetically organized, sometimes within families of functions.
Network Administrator's Guide	Administrators, technical support personnel	Network setup and maintenance for a product
Printer Guide	All users	Information about how to use various printers with the product.
Programmer's Guide	Programmers	Programming concepts and techniques.
Programmer's Reference	Programmers	Technical information, usually about modifying, not using, a product. Includes information about application programming interfaces.
Quick Reference	All users	Brief, concise information about commands or features. Avoid using *Guide* or *Pocket Guide* as part of the title.
Resource Kit	Engineers, technicians, support staff, administrators	Overview of a product's technical features and underpinnings.

(continued)

Title	Audience	Content
Road Map	All users	A learning path or a guide to the printed and/or online product documentation.
Switching to [Microsoft Product]	Users of another, similar product	Information for users of a similar product about how to easily learn the Microsoft product. Often maps features and commands from one to the other.
Technical Reference	Product developers and technical end users, not necessarily programmers	Similar to a *Programmer's Reference,* but does not include information about application programming interfaces. Often covers customizing end-user software.
User's Guide	End users	Information about installing and using the product, possibly including descriptions of new features. Possibly the only printed book in the package.

toggle (adj)

Use as an adjective, as in *toggle key.* A toggle key turns a particular mode on or off.

Do not use *toggle* as a verb; instead, use *switch, click,* or *turn on* and *turn off* to describe the action. For example, use the specific name of a toggle key or command to refer to what the user should do to switch between modes.

Correct

Use the CAPS LOCK key to switch from typing in capital letters to typing in lowercase.
To turn the Ruler on or off, click Ruler on the Edit menu.

Incorrect

Toggle the CAPS LOCK key on or off to switch from capital letters to lowercase.
To turn the Ruler on or off, toggle Ruler on the Edit menu.

tone

Avoid; use *beep* as a noun to refer to a sound, as in "when you hear the beep," unless the user can choose a particular sound.

Tone of Documentation

The tone or writing style of documentation can vary, depending on the product, its audience, and the writer's voice. The most often used tone and the one best suited to most documentation, however, is what is often called "general English." It falls in the middle of the spectrum ranging from informal to formal written English.

A B C D E F G H I J K L M N O P Q R S T U V W X Y Z

General English

General English follows standard grammatical conventions. Sentence length varies, but seldom are sentences longer than 25 words. Most sentences are simple or compound. Complex and compound-complex sentences appear infrequently. The vocabulary favors common straightforward words rather than more obscure words or nominalizations. (A nominalization is a noun formed from a verb, used in a phrase such as "make reference to" rather than "refer to.")

Most books and general interest magazines are written in general English. Most documentation should be written in general English as well. In the examples, the one in general English (the actual example) is clearly the most appropriate.

General English example

Some people who browse Web pages turn off the display of graphics and videos so they can browse the World Wide Web more quickly.

Formal English

Formal English uses more long complex sentences and complete grammatical constructions. Passive voice appears often, as do technical terms and nominalizations. Formal English may be appropriate for some technical material, but if the same sense can be conveyed by simpler general English, it should be. Sometimes a writer can lose track of correct grammar (say of subject-verb agreement) in convoluted formal English sentences.

Formal English example

Some people who browse through World Wide Web pages turn off the display of multimedia such as graphics and videos, which enables them to browse the Web more quickly.

Informal English

Informal English is breezy and journalistic. It's marked by short sentences, contractions, and directly addressing the user. Some grammatical constructions may not be complete. Informal English may be appropriate for some consumer products, but some attempts to be friendly can sound condescending.

Informal English example

You can turn off graphics to surf the Web more quickly.

SEE ALSO **International Considerations**

toolbar

Toolbar is the standard term for the various rows below the menu bar that contain buttons and commands for commonly used tasks. Do not use *command bar* except in technical documentation.

Toolbar

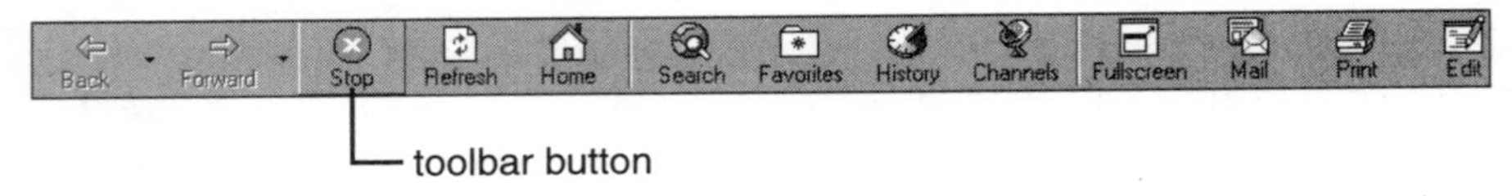

The main toolbar that is on by default and contains buttons for basic tasks, such as opening and printing a file, is called the "standard toolbar." Some programs also have a formatting toolbar that is on by default. Other toolbars can be turned on, usually by clicking Toolbars on the View menu.

In procedures, do not mention the name of a toolbar that is on by default.

Correct

1 Select the words you want to format as bold, and then click Bold.
2 On the Drawing toolbar, click Insert WordArt.

Incorrect

1 On the standard toolbar, click Bold.

toolbox

Generically, a toolbox is a collection of drawing or interface tools such as paintbrushes, lines, circles, scissors, and spray cans. In programming applications such as Visual Basic, the toolbox also includes controls users can add to programs, such as command buttons and option buttons. Tools in a toolbox differ from the commands on a toolbar in that the shapes or controls often can be dragged to a document and manipulated in some way.

Treat elements in a toolbox like any other options in dialog boxes. That is, in procedures tell users to click a particular option, and use bold for toolbox labels. Do not capitalize *toolbox* except to match the interface or if it's a specifically named product feature.

Correct

Insert a **Combo Box** control in the dialog box.

Toolbox

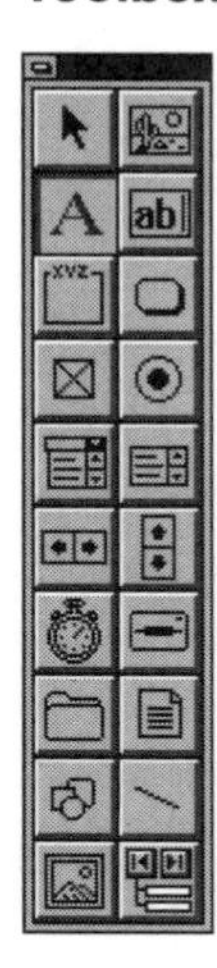

SEE ALSO **Dialog Boxes and Property Sheets**

A B C D E F G H I J K L M N O P Q R S T U V W X Y Z

toolkit

One word.

ToolTip

One word, with both *T*s capitalized, to refer to the feature. ToolTip is not a trademark.

SEE ALSO **ScreenTip**

top left, top right

Avoid; use *upper left* and *upper right* instead.

topic

As in *Help topic;* do not use *article* or *entry.*

toward

Not *towards.*

trackball

A stationary input device that holds a ball, which is moved with the fingers or palm in order to move the pointer. It is acceptable to use *input devices* when referring generically to pens, mice, trackballs, styluses, and so on.

Trademarks

A trademark is a word, name, device, or phrase that has been used to identify goods or services made or provided by a company and to distinguish them from those produced or provided by others. A trade name is the name under which a company does business. Microsoft is a *trade name* when it refers to the company (that is, Microsoft Corporation) and a *trademark* when it represents the brand name, which refers to a product or product line (for example, Microsoft® Word).

A company must protect its trademark rights to ensure that its trademarks continue to represent high standards to its customers and competitors. Trademark rights are based on the continuous and proper use of the mark. Properly used, rights in a trademark can be protected indefinitely.

There are two parts to indicating trademarks; use both in documentation:

- Trademark symbols (™ and ®), also known as "bugs"
- Trademark footnotes, statements that indicate trademark status and ownership

These are discussed in the following sections.

Use of the ™ and ® symbols with Microsoft products

If Microsoft claims rights to a trademark that is not yet registered in the U.S. Patent and Trademark Office, the trademark must be identified by the ™ symbol. After the mark has been registered, it must be identified by the ® symbol.

Correct

Microsoft® Active Desktop™

- Because legally a trademark is an adjective, Microsoft trademarks should not be used as verbs or nouns or in the possessive or plural form.

  Correct

  A case of Microsoft® PowerPoint® presentation graphics programs
  The Microsoft® CodeView® debugger's interface

  Incorrect

  A case of Microsoft PowerPoints®
  CodeView's® interface
  CodeView®'s interface

- Use the name of the product, its descriptor, and, if applicable, its platform, trademarked correctly, on the title page and at the first mention in text. From then on use the descriptor 50 percent of the time.

  Correct

  Microsoft® Windows® operating system
  Microsoft® Visual C++® development system for Microsoft® Windows®

- In printed documentation and CD-ROM books (that is, the complete product documentation presented online and taking the place of printed books), the ™ and ® symbols should be used at the first mention of a Microsoft trademark on the cover or title page and the first mention in the main text.
- Use trademark symbols at the first mention of Microsoft trademarked products on each Web page, or at first mention in a multipage topic or article whose pages are designed to be viewed sequentially.
- Do not use trademark symbols in Help or other software elements. Trademark bugs appear only in the Help About box. The footnotes appear on the product box or the copyright page of a book in the product.
- If the first mention of a product occurs in a heading, it is not necessary to use the trademark symbol, provided that the trademarked product is mentioned immediately in the text that follows the heading.
- Use a registered trademark symbol (®) for the name Microsoft only when it is used as a trademark, not as a trade name. That is, Microsoft should get a trademark symbol when it appears before a product name or a product line, but not when it refers to the corporation.

A B C D E F G H I J K L M N O P Q R S T U V W X Y Z

Correct
Microsoft® Word
Microsoft® development tools
Microsoft Corporation

Incorrect
Microsoft® employee

- As much as possible, "Microsoft" should precede each product name at first mention and be bugged at first mention *per product*.

  Correct
  Planning a trip? Use the Microsoft® Expedia™ travel service to book your flight, and then use the Microsoft® Sidewalk™ city guide to get around town.

- Some products include the name Microsoft as part of a trademarked unit. In these instances, the trademark symbol should follow the entire unit.

  Correct
  Microsoft Press®

  Windows NT® 4.0 Workstation

  Incorrect
  Microsoft® Press

  Windows® NT 4.0 Workstation

- Do not use trademark symbols on the copyright page or screen (Help About), in the table of contents, in the index of printed documentation, or on the copyright screen of a CD-ROM book.

Use of trademark footnotes with Microsoft products

The second step in indicating trademarks is the use of trademark footnotes, statements that indicate trademark status and ownership. These usually appear on the copyright page.

- Do not use trademark symbols in footnotes.
- Trademark footnotes should always be complete sentences that end with a period. List "Microsoft" as a trademark first, then other Microsoft trademarks, then the trademarks licensed to Microsoft, and finally the trademarks of all others, in alphabetical order by company name.
- Do not use Microsoft trademark footnotes either on the startup screen or in Help About, unless there is no other product documentation.
- Do not use Microsoft trademark footnotes on Web pages. Instead, link to the generic Microsoft copyright and trademark notice at //www.microsoft.com/misc/cpyright.htm.

Correct Microsoft trademark footnote

Microsoft, Bookshelf, Visual C++, Windows, and Windows NT are either registered trademarks or trademarks of Microsoft Corporation in the U.S.A. and/or other countries.

SEE ALSO **Copyright Pages and Copyright Screens**

trailing (adj)

Acceptable to mean "following," as in *trailing period, trailing slash,* or *trailing space.*

trash

Refer to the Macintosh icon as *the Trash.*

TRUE

Use all caps as a return value in programming documents.

turn on, turn off

Do not use to refer to selecting or clearing check boxes in procedures. Use *select* and *clear* or *click to select* instead. It is acceptable to use in text to refer to the status options such as multimedia on Web pages (as in "you can turn graphics off").

Use instead of *power on, power off, start,* or *switch on, switch off* to mean turning on and off the computer. Do not separate the two words. That is, do not use a phrase such as *turn power on.*

turnkey (n, adj)

Always one word.

tutorial

Not *CBT* (for *computer-based training).* Use *online tutorial* to distinguish from a printed tutorial if necessary.

two-dimensional, 2-D (adj)

Two-dimensional is preferred, but *2-D* is acceptable. Spell out at first mention. Use *2-D* in tables and indexes and where space is a problem, as well as to reflect the interface.

Hyphenate both the spelled out and abbreviated versions.

type vs. enter

Use *type,* not *type in* or *enter,* if information the user types appears on the screen. An exception to this rule is that you can tell users to "enter" a file name, for example, in a combo box when they have the choice of typing a name or selecting one from a list. You can also use a combination of words such as "type or select" if space is not an issue.

Correct

Type your password.
Enter the file name.
Type the path to the server or select it from the list.

Incorrect

Type in your password.
Enter your password.

SEE ALSO **press**

U.K. (n, adj)

Acceptable abbreviation for *United Kingdom* as a noun or adjective, but avoid except in tables or to save space. If you use the abbreviation, it's not necessary to spell out at first mention. Always use periods and no space.

un (prefix)

In general, do not hyphenate words beginning with *un,* such as *undo* and *unread,* unless it's necessary to avoid confusion, as in *un-ionized,* or unless *un* is followed by a proper noun, as in *un-American.* If in doubt, check *American Heritage Dictionary* or your project style sheet.

unavailable

Use *unavailable,* not *grayed* or *disabled,* to refer to unusable commands and options on the interface. Use *dimmed* only if you have to describe their appearance.

Correct

You cannot use unavailable commands until your file meets certain conditions, such as having selected text. These commands appear dimmed on the menu.

SEE ALSO **dimmed, disable, Menus and Commands**

UNC

Acronym for "Universal Naming Convention," the system for indicating names of servers and computers, such as "\\Servername\Sharename." Use only in technical documentation.

uncheck, unmark, unselect

Do not use for check boxes or selections; use *clear the check box* or *cancel the selection* instead.

undelete

Do not use except to reflect the interface; use *restore* instead.

A B C D E F G H I J K L M N O P Q R S T U V W X Y Z

underline, underscore

Use *underline* to refer to text formatting with underlined characters or to formatting. Use *underscore* to refer to the underscore character (_).

undo

Do not use the command name *Undos* as a noun to refer to multiple instances of undoing actions. Write around instead, as in "to undo multiple actions" or "select the actions that you want to undo." It is acceptable to say that a command *undoes* an action.

uninstall (v)

Avoid if possible, especially in end-user documentation; use *remove* instead. However, you can use it as necessary to match the interface or, in programmer documentation, to refer to a particular type of program.

unnamed buttons

If you refer to unnamed buttons that appear in the interface, use the syntax given in the example and insert a bitmap showing the button, if possible.

Correct

Click the Minimize button

Most buttons are named in ToolTips, so if it's not possible to use inline graphics, use the name only.

SEE ALSO **Screen Terminology**

unprintable

Do not use; use *nonprinting* instead.

update (v)

Use instead of *refresh* to describe the action of an image being restored on the screen or data in a table being updated.

Correct

To update the appearance of your screen, click Refresh.

Do not use to refer to product upgrades.

SEE ALSO **refresh**

upgrade

Use instead of *update* to refer to product upgrades.

upper left (n), upper right (n)

Use instead of *top left* and *top right.* Hyphenate as adjectives: *upper-left* and *upper-right.*

uppercase (adj, n)

One word. Do not use *uppercased.* Avoid using as a verb.

uppercase and lowercase

Not *upper- and lowercase.*

upsize

Avoid if possible, even though it's becoming more common in client/server products and their terminology. It's jargon for *scale up.*

upward

Not *upwards.*

URL, address

A Uniform (not "Universal") Resource Locator (URL) is an Internet address that locates a specific resource on the World Wide Web or elsewhere on the Internet. It consists of the Internet protocol name; a host name; and optionally other elements such as a port, directory, and file name. Each main element is lowercase (unless case sensitive).

Typical URLs separate the protocol name (such as *http:*) from the rest of the destination with two forward slashes and separate the host and other main elements from each other with one forward slash.

Unless the Internet protocol is not HTTP —such as FTP or GOPHER—it is unnecessary to include the protocol connection at the beginning of the URL. Most browsers assume the connection is through HTTP and add it automatically. Likewise, in straight text, marketing materials, and the like, it's not necessary to include the forward slashes.

However, in procedures or when telling users to type a particular URL, do include the protocol name and forward slashes to avoid any confusion. If possible, set the URL off from the main text to avoid including end punctuation.

It is a good idea to include the trailing slash in a URL unless the final element is a file name, such as "filename.htm." It ensures that the server will return a Web site instead of cycling back to the browser to confirm that the request was not for a browsable file.

Typical URLs

http://www.slate.com/cover/current/cover.asp
http://www.microsoft.com/
www.microsoft.com
ftp://ftp.isbiel.ch/udd/ftp/windows.95/

In general, use *address* to refer to a URL, especially in end-user documentation. Use *URL* when referring to technical aspects of the Internet or in technical documentation.

Use the preposition "at" with the location of an address:

Correct

You can find information about Microsoft products at
www.microsoft.com.

Other URLs

E-mail and bulletin board addresses are also URLs. Their format differs from those in the previous description:

Bulletin board URLS

clari.news.women
news.announce.newusers

U.S. (adj)

Acceptable abbreviation for *United States* only when used as an adjective; do not use *US, USA,* or *U.S.A.* Avoid except in tables or to save space. If you use the abbreviation, it's not necessary to spell out at first mention. Always use periods and no space. Spell out *United States* as a noun except when third-party legalese specifies otherwise.

usable

Not *useable.*

Usenet

The collection of computers and networks that share news articles. Overlaps with the Internet, but not identical to it. From "User Network"; sometimes seen all capped.

user name

Two words unless reflecting the interface.

using vs. with

Avoid *with* to mean "by using"; it is ambiguous and makes localization more difficult.

Correct

You can select part of the picture by using the dotted rectangle selection tool.

Incorrect

You can select part of the picture with the dotted rectangle selection tool.

With is acceptable in some marketing materials and sometimes with product names.

Correct

With Home Essentials, you can create professional documents quickly and easily.

utility, utility program

Not *external command.*

utilize

Do not use; use *use* or another appropriate synonym instead. *Utilize* means using something in a way in which it was not meant to be used, but in technical and business jargon it has become a synonym for the clearer and more concise *use.*

Correct

Some applications are unable to use expanded memory.
If a form contains many fields that use the same information, you can repeat the information with the ASK and REF fields.

value axis

In spreadsheet programs, refers to the (usually) vertical axis in charts and graphs that shows the values being measured or compared. For clarity, refer to it as the "value (y) axis" at first mention; "y axis" is acceptable for subsequent mentions. You can also use "vertical (y) axis" in documentation for novices.

SEE ALSO **category axis**

VCR

Abbreviation for *video cassette recorder.* It is not necessary to spell out the term at first use.

Verbs

Simple verbs in the present tense are easier to read and understand than complex-tense verbs, such as verbs in the progressive, perfect, past, or future tense. Use the active voice, which is more forceful and clearer than the passive voice.

Follow these guidelines for using verbs most effectively:

- Avoid weak, vague verbs such as *be, has, make,* and *do.* Use direct, active verbs instead. *Be* in particular lends itself to the passive voice.

  Stronger
  Windows 98 includes many colorful screen savers.
  Windows provides many multimedia features.
  You can create a new folder.
  Back up your files as part of your regular routine.

  Weaker
  Several colorful screen savers are in Windows 98.
  Windows has many multimedia features.
  You can make a new folder.
  Do a backup of your files as part of your regular routine.

- Do not turn verbs into nouns or create verbs from nouns. Rephrasing frequently makes the text clearer and less awkward.

  Correct
  You can search the document for this text and replace it.
  You can use the Paste Link command to place the data into the worksheet.

  Incorrect
  You can do a search-and-replace on the document.
  You can paste link the data into the worksheet.

- Do not use transitive verbs as intransitive verbs. A transitive verb takes a direct object, which is the receiver of the action of the verb. An intransitive verb does not have a direct object. Do not use the following transitive verbs in active constructions without objects: *complete, configure, display, install, print,* and *process.*

  Correct
  The screen displays information. [transitive]
  A dialog box appears on the screen. [intransitive]
  Restarting Windows completes Setup.

  Incorrect
  A dialog box displays.
  After you restart Windows, Setup completes.

- You can use a transitive verb with a passive construction.

  Correct
  A dialog box is displayed on the screen.
  After you restart Windows, Setup is complete.

For information about verbs in procedures and commands, see **Procedures**.

SEE ALSO **Active Voice vs. Passive Voice, Contractions, Mood of Verbs, tense**

version numbers

Microsoft products can indicate a version in one of two ways: by year of release (becoming more common) or by chronological version number.

By year of release

Many new products have date designations that supersede version numbers—Windows 95, for example. Use these designations as part of the product name when you specify the environment, in the copyright notice on the Help About screen, and instead of version numbers if you must specify product compatibility.

Correct

Microsoft Word for Windows 95
Microsoft Outlook 97
Microsoft Office for Windows 95

Incorrect

Microsoft Word for Windows 95 version 7.0
Microsoft Word version 7.0 for Windows 95
Microsoft Word 7.0 for Windows 4.0

By version number

A complete product version number has several components:

- Version number: *X.xx*
- Release number: *x.Xx*
- Bug-fix number: *x.xX*

Usually, only the version and release numbers are significant to the user. Bug fixes do not introduce new features as do new releases and new versions of a product; they seldom require documentation.

Correct

Microsoft Windows NT version 5.0
Microsoft Internet Explorer 4.0
Microsoft Exchange Server 4.0.829

General guidelines

As a rule, avoid specifying the date designation or version number of a product within the body of text. You may need to specify the version number, however, when comparing current and previous versions or for reasons of clarity and technical accuracy. Do not specify the bug-fix number of a product unless it's technically relevant.

If you must mention a version number, specify it on the first mention in a topic or section. Thereafter, refer only to the product name without the date designation or version number. Use *.x* (italic) to indicate all release numbers of a product (for example, "Windows NT Server version 3.*x*"), or use *earlier* or *later* (for example, "Windows NT version 3.0 or later"). Do not use *higher* or *lower.*

It is acceptable to append the version number directly to the product name (as in "Windows NT Server 3.5") after the product name has been introduced at first mention with any trademark descriptor plus the word *version.*

Correct

If you are using Windows NT Server version 3.5 or later, ... [first mention]
If you are using Windows NT Server 3.5 or later, ... [after first mention]
If you are using Microsoft Excel for Windows 95, ... [first mention]
You can use Microsoft Excel to. … [after first mention]

versus, vs.

In headings, use the abbreviation *vs.,* all lowercase. In text, spell out as *versus.*

Correct

Daily vs. Weekly Backups

A B C D E F G H I J K L M N O P Q R S T U V W X Y Z

VGA, SVGA

Do not use *VGA+* to describe video adapters of higher resolution than VGA; there is no such thing. Use *Super VGA (SVGA)* instead. Spell out *Super* (but not *video graphics adapter*) at first mention.

via

Via implies a geographic context. Avoid using *via* as a synonym for *by, through,* or *by means of.* Use the most specific term instead.

video adapter

Use to refer to the expansion board that converts images to the electronic signals needed by the monitor. Do not use *video graphics board* unless you specifically mean something that accepts video images from a camera or VCR. Likewise, do not use the term *card, board,* or *controller* because the specific hardware varies.

video board

Do not use. Use *video adapter* or *video graphics board,* depending on your specific meaning.

video display

Do not use; use *screen* instead. See **screen**.

viewport

One word. Refers to a view of a document or image in computer graphics programs.

virtual (adj)

Technical term used to describe a device or service that appears to the user as something it actually is not or that doesn't physically exist. For example, a virtual disk performs like a disk but is actually a part of the computer's memory. Some other virtual devices or services are virtual machine, virtual memory, and virtual desktop. Use the term only to refer to the specific element.

Avoid in end-user documentation.

For more information, see *Microsoft Press Computer Dictionary.*

virtual root

Acceptable to use to refer to the root directory that the user sees when connected to an Internet server. It is actually a pointer to the actual root directory. Do not use *virtual directory* as a synonym.

virtual server

Acceptable in technical documentation to refer to a server that looks like a physical server to a browser. Sometimes used as a synonym for "Web site." Use *Web site* instead if possible in the context.

virtualize

Avoid. Use sparingly if you must, and only in technical documents. Try to use phrasing such as "access virtual memory" instead.

voice

In general, use the active voice. See **Active Voice vs. Passive Voice**, **Mood of Verbs**, and **Verbs**.

voice mail

Two words. Do not abbreviate as *v-mail* or *vmail*.

W3C

Abbreviation for World Wide Web Consortium, the organization that sets standards for the Web and HTML. Okay to use, but define at first mention.

want

Not *wish.* Do not confuse with *need.* Be sure to use the term that is appropriate to the situation. *Need* connotes a requirement or obligation; *want* indicates that the user has a choice of actions.

Correct

If you want to use a laser printer, you need a laser printer driver.

Incorrect

If you wish to format the entire word, double-click it.

Warnings

A warning advises users that failure to take or avoid a specific action could result in physical harm to the user or the hardware. Use a warning, not a caution, when such damage is possible.

SEE ALSO **Notes and Tips**

we

In general, do not use, except in some marketing or installation information, where *we recommend* is acceptable. Otherwise, *we* can sound patronizing.

SEE ALSO **First Person, Second Person**

Web

Okay to use to refer to the World Wide Web. For novice users, always use "World Wide Web" first, then shorten to "Web."

Always capitalize when used alone. A few words formed from *Web,* such as "webzine," are usually not capitalized, but if in doubt, capitalize.

A
B
C
D
E
F
G
H
I
J
K
L
M
N
O
P
Q
R
S
T
U
V
W
X
Y
Z

Weblication

Jargon for "Web application." Do not use. It could be confused with "Web publication."

where

Use to introduce a list, as in code or formulas, to define the meaning of elements such as variables or symbols.

Correct

Use the following formula to calculate the return, where:
r = rate of interest
n = number of months
p = principal

while

Use to refer to something occurring in time. Avoid as a synonym for *although,* which can be ambiguous.

Correct

Fill out your registration card while you wait for Setup to be completed.
Although the icon indicates that the print job is finished, you may have to wait until a previous job is finished.

Incorrect

While the icon indicates that the print job is finished, you may have to wait until a previous job is finished.

white paper

Always two words.

white space (n), white-space (adj)

Two words as a noun, hyphenated as an adjective.

who vs. that

Although there is no linguistic basis for not using *that* to refer to people, as in "the man that was walking," it's considered more polite to use *who* instead of *that* in references to people. Therefore, use *who,* not *that,* to introduce clauses referring to users.

Correct

Custom Setup is for experienced users who want to alter the standard Windows configuration.

Incorrect

Custom Setup is for experienced users that want to alter the standard Windows configuration.

wildcard character

Always use the word *character* with *wildcard* when referring to a keyboard character that can be used to represent one or many characters, such as the * or ?

Wildcard is one word.

window (n)

Do not use as a verb.

Windows, Windows-based

Use *Windows* as a modifier for aspects or elements of the Windows operating system itself. Because Windows is a Microsoft registered trademark, do not use *Windows* to modify programs, hardware, or development methods that are based on or run on the Windows operating system. Instead, use *Windows-based* or *running Windows*. To avoid a ridiculous construction, the term *Windows user* is acceptable.

Correct

Windows-based program
Windows-based device
the Windows Recycle Bin
a computer running Windows

Incorrect

Windows program
Windows computer

Windows Explorer

A program in Windows 95 that shows the hierarchical structure of all the folders on a computer and the files and folders in selected folders. It replaces File Manager. Do not precede with *the* and do not shorten to *Explorer*.

Windows NT Server, Windows NT Workstation

Avoid the use of *Windows NT* as a modifier for aspects or elements of the Windows NT Server or Windows NT Workstation products. Instead, say that a product or process *runs on* one of these products, not *is on* one of them.

Correct

The printer is attached to a computer running Windows NT Server.
The Microsoft Exchange Client software is on a computer running Windows NT Workstation.

Incorrect

Windows NT Server server
Windows NT–based program
Windows NT system

A B C D E F G H I J K L M N O P Q R S T U V W X Y Z

Winsock

Okay to use to refer to the Windows Sockets API. Avoid *Sockets.*

wireframe

One word. Refers to a type of 3-D graphic.

wizard

Always use lowercase for the generic term *wizard.* Capitalize *wizard* if it's part of a feature name that appears in the interface. The names of feature wizards such as Internet Connection Wizard should always be checked with the legal department to be sure they are not trademarked by another company and to determine whether Microsoft should pursue registering them. Currently, the only trademarked Microsoft wizard is TipWizard.

word processing

Hyphenate word-processing words according to part of speech, as shown:

- Word-processed (adj), word-processing (adj)
- Word processor (n), word processing (n)

Avoid using *word process* as a verb. Use *write, format,* or another more specific term instead.

wordwrap

One word. It's acceptable to use *wordwrapping,* as in "turn off wordwrapping."

work area

Do not use unless it's a product-specific term, as in Visual FoxPro. Instead, use *desktop* for the screen area not contained within the workspace.

workgroup

Workgroup is one word. When referring to collaboration software, *groupware* is acceptable, but use *workgroup software,* if possible, to refer to such programs. For Microsoft workgroup software, however, try to use the actual product descriptor—for example, *messaging system* for Microsoft Exchange Server.

working memory

Do not use; use *available memory* instead.

worksheet

Do not use as a synonym for *spreadsheet.* A spreadsheet is a computer accounting program, such as Microsoft Excel; a worksheet is the document produced by a spreadsheet program.

workspace

One word. Refers to the area within the application window. Use *client area* only if necessary in technical documentation.

workstation

One word. Use to refer to a personal computer used by an individual in a network. It's the client in a client/server system.

World Wide Web

The graphical system of moving through the Internet using hyperlinks. Documents are formatted using Hypertext Markup Language (HTML). These documents are on Web servers that use the Hypertext Transfer Protocol (HTTP) to deliver the Web pages.

Spell out at first mention in material for novices, then use "the Web." Capitalize in all instances, including references to internal Web sites (intranets).

Use "on" to refer to material existing on the Web. You can use "to" or "on" to refer to the action of creating and publishing something "to the Web" or "on the Web."

write-only (adj)

Always hyphenated. Related to *read/write,* but *write-only* and *read-only* refer to properties of files, and *read/write* refers to a level of permissions granted to users, not an adjective defining files or other objects.

write-protect (v), write-protected (adj)

Always hyphenated. Use *write-protect,* not *lock,* to refer to the action of protecting disks from being overwritten.

Correct

to write-protect a disk
a write-protected disk
a disk that is write-protected

WWW

Abbreviation for World Wide Web. Okay to use if it appears in the interface, but in general use *the Web* instead.

SEE ALSO **World Wide Web**

x

Use a lowercase italic *x* as a placeholder number or variable. Do not use to refer to a generic unspecified number; use *n* instead.

Correct

version 4.*x*
R4*x*00

Incorrect

Move the insertion point *x* spaces to the right.

However, do not use 80*x*86. Refer to the specific processor; if necessary, it's okay to say something like "80386 (or *386*) or higher."

SEE ALSO ***n***

x-axis

In general, use *category (x) axis* to refer to the (usually) horizontal axis in charts and graphs that shows the categories being compared. Include a reference to the horizontal axis if it will clarify the meaning. Note lowercase, hyphen, and roman.

Xbase

Not *xBASE.*

x-coordinate

Note lowercase, hyphen, and roman.

XON/XOF

Note all uppercase letters and slash mark. Refers to the handshake between two computers during transmission of data.

y-axis

In general, use *value (y) axis* to refer to the (usually) vertical axis in charts and graphs that shows the values being measured or compared. Include a reference to the vertical axis if it will clarify the meaning. Note lowercase, hyphen, and roman.

y-coordinate

Note lowercase, hyphen, and roman.

Z

z-

Hyphenate all words referring to entities that begin with *z* used as a separate letter, such as *z-axis, z-coordinate, z-order,* and *z-test.* Follow your project's style sheet for capitalization.

z-axis

In 3-D charts, the z-axis shows depth. It generally represents values. Refer to the z-axis as the *value axis,* where both the x-axis and y-axis are category axes, but include *z-axis* in parentheses if it will clarify the meaning. Note lowercase, hyphen, and roman.

zero (s), zeros (pl)

Not *zeroes.*

In measurements, when the unit of measurement is not abbreviated, zero takes the plural, as in "0 megabytes."

zero character

In the ASCII character set, a zero character represents the digit 0 but is ASCII code 48. Differentiate it from the NUL character (ASCII code 0).

SEE ALSO **NUL, null, NULL, Null**

zero-length string

A string that contains only null characters or whose length is zero. Also called *null string*.

SEE ALSO **NUL, null, NULL, Null**

z-order

Refers to the visual layering of objects, such as windows, on the screen. The term refers to the layering along the z-axis, which shows depth.

ZIP Code

Avoid in general text; instead, use the generically international *postal code.* In forms or fill-in fields include *Postal Code* as well as ZIP Code. It is acceptable to combine the two terms to save space.

Correct

ZIP Code/Postal Code:______
ZIP/Postal Code:______
Postal Code:______

NOTE Use an uppercase *C* for *Code* in accordance with U.S. Postal Service style. Also, ZIP Code is a registered trademark.

zoom box

Macintosh screen element that serves essentially the same purpose as the Maximize and Minimize buttons do in the Windows interface. See **Screen Terminology**.

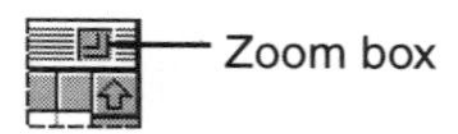

zoom in, zoom out

Do not use *dezoom* or *unzoom.*

Appendix A

List of Acronyms and Abbreviations

Terms are listed alphabetically by acronym or abbreviation, not by meaning. Unless otherwise noted in the "Comments" column, all terms are acceptable to use after spelling out at first mention. Some terms are never acceptable, and some need not be spelled out. Place the spelled-out meaning in parentheses immediately following the acronym. When spelling out the term's meaning, follow the capitalization as given in the "Meaning" column unless noted otherwise.

Exclusion of a term does not mean it is not acceptable. Many product-specific terms do not appear here. Follow your project style sheet.

Acronym	*Meaning*	*Comments*
ACL	access control list	
ADK	application development kit	
ADO	ActiveX Data Objects	Not *Active Data Objects.*
ADSL	asymmetric digital subscriber line	
ADT	application development toolkit	Avoid; use *ADK* instead, if possible.
ANSI	American National Standards Institute	
API	application programming interface	Don't use *application program interface.*
APPC	Advanced Program-to-Program Communications	A protocol in Systems Network Architecture.
ASCII	American Standard Code for Information Interchange	Not necessary to spell out at first mention.
ASP	Active Server Pages	A product and file type for building applications for Internet Information Server. Use the phrase *ASP page.*
ATM	asynchronous transfer mode	
A/V	audio/video	
AVI	Audio-Video Interleaved	Microsoft's multimedia file format for Windows.

Acronym	*Meaning*	*Comments*
BASCOM	Basic Compiler	Do not use acronym.
Basic	Beginners All-purpose Symbolic Instruction Code	Initial cap only. Do not spell out.
BBS	bulletin board system	
BFTP	Broadcast File Transfer Protocol	
BID	board interface driver	
BIFF	Binary Interchange File Format	
BIOS	basic input/output system	
BISYNC	Binary Synchronous Communications Protocol	
BSC	Binary Synchronous Communications Protocol	Use *BISYNC* instead.
BSMS	billing and subscriber management system	
CBT	computer-based training	Avoid; use *tutorial*.
CD	compact disc	See **CD-ROM, CD-ROM drive**.
CDS	Circuit Data Services	
CD-ROM	compact disc read-only memory	See **CD-ROM, CD-ROM drive**.
CGA	color/graphics adapter	
CGI	Common Gateway Interface	
CIS	computer information systems	
CISC	complex instruction set computer	Not necessary to spell out in technical documentation.
CMC	1. Continuous Media Controller 2. Common Messaging Calls (MAPI term)	
CMOS	complementary metal oxide semi-conductor	
CMS	continuous media server	Do not use; use *MMS* instead.
CMY	cyan-magenta-yellow	
CMYK	cyan-magenta-yellow-black	
COM	Component Object Model	
CPI-C	Common Programming Interface for Communications	
CPU	central processing unit	Do not spell out.
CR/LF	carriage return/line feed	

(continued)

Acronym	*Meaning*	*Comments*
CRT	cathode-ray tube	Do not spell out.
CSR	customer service representative	
CTI	Computer-Telephony Integration	
DAE	data access engine	
DAO	Data Access Object	
DAT	digital audio tape	
DBCS	double-byte character set	
DBMS	database management system	
DCE	distributed computing environment	
DCOM	distributed version of Component Object Model	Avoid; use *COM* instead. See **COM, ActiveX, and OLE Terminology**.
DDBMS	distributed database management system	
DDE	Dynamic Data Exchange	
DDI	Device Driver Interface	
DDL	data definition language	
DES	Data Encryption Standard	
Dfs	distributed file system	Spell as shown.
DHCP	Dynamic Host Configuration Protocol	
DIB	device-independent bitmap	
DIF	Data Interchange Format	
DLC	Data Link Control	
DLL	dynamic-link library	Do not use *dynalink*.
DMOD	dynamic address module	
DNA	[Windows] Distributed interNet Applications [Architecture]	This spelling is required for legal and tradename reasons.
DNS	1. Domain Name System 2. Domain Name Server	The Domain Name Server implements the Domain Name System. Differentiate between the two.
DOS	disk operating system	Do not spell out. Avoid except as *MS-DOS*.
DSP	digital signal processor	
DSS	1. decision support system 2. digital satellite system	
DVD	digital video disc	Do not spell out as *digital versatile disc*.

Acronym	Meaning	Comments
EA	extended attributes	Do not use abbreviation.
ECC	electronic credit card	
EISA	Extended Industry Standard Architecture	Not necessary to spell out in technical documentation.
EGA	enhanced graphics adapter	
EPS	encapsulated PostScript	
FAQ	frequently asked questions	Precede with *a*, not *an*.
FAT	file allocation table	
FAX	facsimile	Do not use *FAX;* use *fax* instead. Do not spell out at first mention.
Fortran	Formula Translation	Initial cap only. Do not spell out.
FTP	File Transfer Protocol	All lowercase when used in an Internet address.
FTS	Financial Transaction Services	
GDI	Graphics Device Interface	
GIF	Graphics Interchange Format	
GPI	graphics programming interface	
GUI	graphical user interface	
GUID	globally unique identifier	
HAL	hardware abstraction layer	
HBA	host bus adapter	
HDLC	High-level Data Link Control	An information transfer protocol.
HMA	high-memory area	
HPFS	high-performance file system	
HROT	host running object table	
HTML	Hypertext Markup Language	Not *HyperText*.
HTTP	Hypertext Transfer Protocol	Not *HyperText*. All lowercase (http) when used in an Internet address.
IBN	interactive broadband network	
ICP	independent content provider	
IDE	1. integrated device electronics 2. integrated development environment	Sometimes seen as *integrated drive electronics*. Spell out at first mention and use one consistently.
IEEE	Institute of Electrical and Electronics Engineers, Inc.	

(continued)

Acronym	*Meaning*	*Comments*
IFS	installable file system	
IHV	independent hardware vendor	
IISAM	installable indexed sequential access method	
I/O	input/output	
IOCTL	I/O control	
IOS	integrated office system	
IP	Internet Protocol	
IPC	interprocess communication	
IPX/SPX	Internetwork Packet Exchange/ Sequenced Packet Exchange	*SPX* also seen as Session Packet Exchange.
IRC	Internet Relay Chat	Service for real-time online conversation.
IS	Information Services	
ISA	Industry Standard Architecture	Not necessary to spell out in technical documentation.
ISAM	indexed sequential access method	
ISAPI	Internet Server Application Programming Interface (or *Internet Server API*)	
ISDN	Integrated Services Digital Network	
ISO	International Standards Organization	
ISV	independent software vendor	
ITV	interactive TV	
JPEG	Joint Photographic Experts Group	Refers both to the standard for storing compressed images and a graphic stored in that format.
LADDR	layered-architecture device driver	
LAN	local area network	
LCD	liquid crystal display	Not necessary to spell out at first mention.
LDTR	local descriptor table register	
LED	light-emitting diode	
LISP	List Processor	
LRPC	lightweight remote procedure call	Automation only.
LU	Logical Unit	End point in an SNA network.
MAC	media access control	Do not use acronym.
MAN	metropolitan area network	

Acronym	*Meaning*	*Comments*
MAPI	Messaging Application Programming Interface	Not necessary to spell out in technical documentation.
MASM	Macro Assembler	
MCA	Micro Channel Architecture	IBM trademark.
MCGA	multicolor graphics array	
MCI	Media Control Interface	
MDA	monochrome display adapter	
MDI	multiple-document interface	
MFC	Microsoft Foundation Classes	
MIDI	Musical Instrument Digital Interface	Not necessary to spell out in technical documentation.
MIDL	Microsoft Interface Definition Language	Do not precede with *the*.
MIF	Management Information Format	
MIME	Multipurpose Internet Mail Extensions	Protocol for defining file attachments for the Web.
MIPS	millions of instructions per second	
MIS	management information systems	Use *IS* instead.
MITV	Microsoft Interactive TV	
MMOSA	Microsoft multimedia operating system architecture	Not necessary to spell out in technical documentation.
MMS	Microsoft Media Server	Do not precede with *the*.
MMU	memory management unit	
MOF	Managed Object Format	
MPEG	Moving Picture Experts Group	Sometimes called "Motion Picture(s) Expert(s) Group, but follow the spelling given.
MS	Microsoft	Do not use as abbreviation.
MSBDN	Microsoft Broadcast Data Network	
MSMQ	Microsoft Message Queue Server	
MSN	The Microsoft Network	
MSO	multiple service operator	
MTA	message transfer agent	
NA	not applicable, not available	Do not use abbreviation, even in tables.
NAC	network adapter card	Do not use abbreviation.

(continued)

Acronym	*Meaning*	*Comments*
NAN	not a number	
NCB	network control block	
NCSA	National Center for Supercomputing Applications	
NDIS	network driver interface specification	
NDK	network development kit	
NetBEUI	NetBIOS Enhanced User Interface	
NetBIOS	network basic input/output system	
NFS	network file system	
NIC	network interface card	
NIK	network integration kit	
NLS	national language support	
NMI	nonmaskable interrupt	
NOS	network operating system	
NT	New Technology	Never spell out. See **Windows NT Server, Windows NT Workstation**.
NTFS	NTFS file system	The redundant phrase is correct. Do not use *NT file system* or *New Technology file system*.
NTSC	National Television System Committee	
OCR	optical character recognition	
ODBC	Open Database Connectivity	
ODL	Object Description Language	Automation only.
ODS	Open Data Services library	
OEM	original equipment manufacturer	Not necessary to spell out in technical documentation.
OIS	office information system	
OLE	object linking and embedding	Do not spell out.
OOFS	object-oriented file system	
OOM	out of memory	Do not use acronym.
OOP	object-oriented programming	
ORPC	object remote procedure call	
OSI	Open Systems Interconnection	
PANS	pretty amazing new stuff (or services)	Refers to telephone services. See **POTS**, page 309.
PAR	Product Assistant Request	Do not use acronym.

Acronym	*Meaning*	*Comments*
PARC	Palo Alto Research Center	
PC	personal computer	Avoid; use *computer* instead.
PCMCIA	Personal Computer Memory Card International Association	
PDC	Primary Domain Controller	
PDF	1. Portable Document Format file 2. Package Definition File	1. File format used by Acrobat. 2. Used in some SDKs.
PERT	program evaluation and review technique	
PFF	Printer File Format	
PIF	program information file	
PIN	personal identification number	
POTS	plain old telephone service	See also **PANS**, page 308.
PPV	pay per view	
PROM	programmable read-only memory	
PSU	power supply unit	
QA	quality assurance	
QBE	query by example	
RAID	1. Redundant Array of Inexpensive Disks 2. retrieval and information database (Microsoft only)	
RAM	random access memory	
RAS	1. remote access server 2. Remote Access Service	Remote Access Service is Microsoft Windows software. The server is a host on a LAN equipped with modems.
RBA	1. relative byte address 2. resource-based approach	
RDBMS	relational database management system	
RFT	revisable form text	
RGB	red-green-blue	Not necessary to spell out in technical documentation.
RIFF	Resource Interchange File Format	
RIP	1. Routing Information Protocol 2. Remote Imaging Protocol 3. Raster Image Processor	Always spell out at first mention to avoid confusion.

(continued)

Acronym	*Meaning*	*Comments*
RIPL	remote initiation program load	
RISC	reduced instruction set computer	Not necessary to spell out in technical documentation.
ROM	read-only memory	
ROM BIOS	read-only memory basic input/output system	
RPC	remote procedure call	
RTF	Rich Text Format	
SAA	Systems Application Architecture	IBM trademark.
SAF	[SQL] Server Administration Facility	
SAMI	Synchronized Accessible Media Interchange	Format used to create a time-synchronized captioning file
SAP	Service Advertising Protocol	
SAPI	Speech API	
SBCS	single-byte character set	
SCSI	small computer system interface	Precede acronym with *a,* not *an* (prounounced “scuzzy”).
SDK	software development kit	
SDLC	synchronous data link control	
SGML	Standard Generalized Markup Language	
SIC	standard industry classification	
SIG	special interest group	
SIMM	single inline memory module	Not necessary to spell out in technical documentation.
SLIP	Serial Line Internet Protocol	Method of transmitting data over serial lines such as phone lines.
SMB	server message block	
SMP	symmetric multiprocessing	
SMS	system management software	
SMTP	Simple Mail Transfer Protocol	
SNA	Systems Network Architecture	
SNMP	Simple Network Management Protocol	
SPI	service provider interface	

Acronym	*Meaning*	*Comments*
SQL	Structured Query Language	See **SQL Server**.
STB	set-top box	Do not use abbreviation.
SVC	switched virtual circuit	
SVGA	super video graphics adapter	Use *Super VGA* when spelled out.
TAPI	Telephony API	
TBD	to be determined	
TCP/IP	Transmission Control Protocol/Internet Protocol	
TIFF	Tagged Image File Format	
TP	transaction processing	
TSPI	Telephony Service Provider Interface	
TSR	terminate-and-stay-resident	
TTY/TDD	teletype/telecommunication device for the deaf	
TV	television	Okay to use without spelling out.
UDP	User Datagram Protocol	
UI	user interface	
UMB	upper memory block	
UNC	universal naming convention	
UPC	universal product code	
UPS	uninterruptible power supply	
URL	Uniform Resource Locator	Sometimes called Universal Resource Locator, but use *Uniform* in Microsoft documents.
UTC	coordinated universal time	Abbreviated from French. Now often called Universal Time Coordinate. Basically the same as Greenwich Mean Time.
UUID	universally unique identifier	
VAR	value-added reseller	
VAT	value-added tax	
VBA	Microsoft Visual Basic for Applications	Do not use acronym. This is a legal requirement to protect Microsoft property.
VCPI	virtual control program interface	

(continued)

Acronym	*Meaning*	*Comments*
VCR	video cassette recorder	Not necessary to spell out at first mention.
VGA	video graphics adapter	
VIO	video input/output	
VM	virtual memory	
VRML	Virtual Reality Modeling Language	
VSAM	virtual storage access method (or memory)	
VSD	vendor-specific driver	
VTP	virtual terminal program	
WAN	wide area network	
WBEM	Web-based Enterprise Management	
WOSA	Windows Open Services Architecture	Not necessary to spell out in technical documentation.
WWW	World Wide Web	Lowercase when used in an Internet address.
WYSIWYG	what you see is what you get	
XCMD	external command	
XML	Extensible Markup Language	
XMS	extended memory specification	
ZAW	Zero Administration for Windows	Collection of Microsoft utility programs.

Appendix B

Special Characters

The following table shows the correct meaning or term to use to describe the special character.

Character	*Name*
´	accent acute
^	accent circumflex, caret
`	accent grave
&	ampersand
< >	angle brackets
’	apostrophe (publishing character)
'	apostrophe (user-typed text)
*	asterisk (not *star*)
@	at sign
\	backslash
{ }	braces (not *curly brackets*)
[ ]	brackets
¢	cent sign
« »	chevrons, opening and closing. Microsoft term, seldom used, especially in documentation. Also referred to as *merge field characters* in Word.
©	copyright symbol
†	dagger
°	degree symbol
÷	division sign
$	dollar sign
[[ ]]	double brackets
...	ellipsis (s), ellipses (pl). Do not add space between ellipsis points.
—	em dash
–	en dash

(continued)

Special Characters

<table>
<tr><th>Character</th><th>Name</th></tr>
<tr><td>=</td><td>equal sign (not equals sign)</td></tr>
<tr><td>!</td><td>exclamation point (not exclamation mark or bang)</td></tr>
<tr><td>></td><td>greater than</td></tr>
<tr><td>≥</td><td>greater than or equal to</td></tr>
<tr><td>-</td><td>hyphen</td></tr>
<tr><td>"</td><td>inch mark</td></tr>
<tr><td><</td><td>less than</td></tr>
<tr><td>≤</td><td>less than or equal to</td></tr>
<tr><td>–</td><td>minus (use en dash)</td></tr>
<tr><td>×</td><td>multiplication sign (use * if necessary to match software)</td></tr>
<tr><td>≠</td><td>not equal to</td></tr>
<tr><td>#</td><td>number sign in most cases, but pound key when referring to the telephone</td></tr>
<tr><td>¶</td><td>paragraph mark or carriage return</td></tr>
<tr><td>( )</td><td>parentheses (pl), opening or closing parenthesis (s)</td></tr>
<tr><td>%</td><td>percent</td></tr>
<tr><td>π</td><td>pi</td></tr>
<tr><td>|</td><td>pipe, vertical bar, or OR symbol</td></tr>
<tr><td>+</td><td>plus</td></tr>
<tr><td>±</td><td>plus or minus</td></tr>
<tr><td>?</td><td>question mark</td></tr>
<tr><td>“ ”</td><td>quotation marks (not quotes or quote marks)</td></tr>
<tr><td>" "</td><td>straight quotation marks (not quotes or quote marks)</td></tr>
<tr><td>‘ ’</td><td>single curly quotation marks (not quotes or quote marks)</td></tr>
<tr><td>' '</td><td>single straight quotation marks (not quotes or quote marks)</td></tr>
<tr><td>®</td><td>registered trademark symbol</td></tr>
<tr><td>§</td><td>section</td></tr>
<tr><td>/</td><td>slash mark (not virgule)</td></tr>
<tr><td>~</td><td>tilde</td></tr>
<tr><td>™</td><td>trademark symbol</td></tr>
<tr><td>_</td><td>underscore</td></tr>
</table>

Index

Special Symbols

A

Index

B

D

E

F

G

J

K

L

M

N

O

P

Q

R

S

T

U

V

W

Z

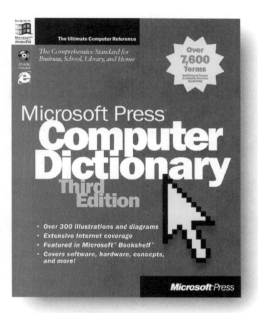

IMPORTANT—READ CAREFULLY BEFORE OPENING SOFTWARE PACKET(S). By opening the sealed packet(s) containing the software, you indicate your acceptance of the following Microsoft License Agreement.

MICROSOFT LICENSE AGREEMENT

(Book Companion CD)

This is a legal agreement between you (either an individual or an entity) and Microsoft Corporation. By opening the sealed software packet(s) you are agreeing to be bound by the terms of this agreement. If you do not agree to the terms of this agreement, promptly return the unopened software packet(s) and any accompanying written materials to the place you obtained them for a full refund.

MICROSOFT SOFTWARE LICENSE

1. GRANT OF LICENSE. Microsoft grants to you the right to use one copy of the Microsoft software program included with this book (the "SOFTWARE") on a single terminal connected to a single computer. The SOFTWARE is in "use" on a computer when it is loaded into the temporary memory (i.e., RAM) or installed into the permanent memory (e.g., hard disk, CD-ROM, or other storage device) of that computer. You may not network the SOFTWARE or otherwise use it on more than one computer or computer terminal at the same time.

2. COPYRIGHT. The SOFTWARE is owned by Microsoft or its suppliers and is protected by United States copyright laws and international treaty provisions. Therefore, you must treat the SOFTWARE like any other copyrighted material (e.g., a book or musical recording) except that you may either (a) make one copy of the SOFTWARE solely for backup or archival purposes, or (b) transfer the SOFTWARE to a single hard disk provided you keep the original solely for backup or archival purposes. You may not copy the written materials accompanying the SOFTWARE.

3. OTHER RESTRICTIONS. You may not rent or lease the SOFTWARE, but you may transfer the SOFTWARE and accompanying written materials on a permanent basis provided you retain no copies and the recipient agrees to the terms of this Agreement. You may not reverse engineer, decompile, or disassemble the SOFTWARE. If the SOFTWARE is an update or has been updated, any transfer must include the most recent update and all prior versions.

4. DUAL MEDIA SOFTWARE. If the SOFTWARE package contains more than one kind of disk (3.5", 5.25", and CD-ROM), then you may use only the disks appropriate for your single-user computer. You may not use the other disks on another computer or loan, rent, lease, or transfer them to another user except as part of the permanent transfer (as provided above) of all SOFTWARE and written materials.

5. SAMPLE CODE. If the SOFTWARE includes Sample Code, then Microsoft grants you a royalty-free right to reproduce and distribute the sample code of the SOFTWARE provided that you: (a) distribute the sample code only in conjunction with and as a part of your software product; (b) do not use Microsoft's or its authors' names, logos, or trademarks to market your software product; (c) include the copyright notice that appears on the SOFTWARE on your product label and as a part of the sign-on message for your software product; and (d) agree to indemnify, hold harmless, and defend Microsoft and its authors from and against any claims or lawsuits, including attorneys' fees, that arise or result from the use or distribution of your software product.

DISCLAIMER OF WARRANTY

THE SOFTWARE (INCLUDING INSTRUCTIONS FOR ITS USE) IS PROVIDED "AS IS" WITHOUT WARRANTY OF ANY KIND. MICROSOFT FURTHER DISCLAIMS ALL IMPLIED WARRANTIES INCLUDING WITHOUT LIMITATION ANY IMPLIED WARRANTIES OF MERCHANTABILITY OR OF FITNESS FOR A PARTICULAR PURPOSE. THE ENTIRE RISK ARISING OUT OF THE USE OR PERFORMANCE OF THE SOFTWARE AND DOCUMENTATION REMAINS WITH YOU.

IN NO EVENT SHALL MICROSOFT, ITS AUTHORS, OR ANYONE ELSE INVOLVED IN THE CREATION, PRODUCTION, OR DELIVERY OF THE SOFTWARE BE LIABLE FOR ANY DAMAGES WHATSOEVER (INCLUDING, WITHOUT LIMITATION, DAMAGES FOR LOSS OF BUSINESS PROFITS, BUSINESS INTERRUPTION, LOSS OF BUSINESS INFORMATION, OR OTHER PECUNIARY LOSS) ARISING OUT OF THE USE OF OR INABILITY TO USE THE SOFTWARE OR DOCUMENTATION, EVEN IF MICROSOFT HAS BEEN ADVISED OF THE POSSIBILITY OF SUCH DAMAGES. BECAUSE SOME STATES/COUNTRIES DO NOT ALLOW THE EXCLUSION OR LIMITATION OF LIABILITY FOR CONSEQUENTIAL OR INCIDENTAL DAMAGES, THE ABOVE LIMITATION MAY NOT APPLY TO YOU.

U.S. GOVERNMENT RESTRICTED RIGHTS

The SOFTWARE and documentation are provided with RESTRICTED RIGHTS. Use, duplication, or disclosure by the Government is subject to restrictions as set forth in subparagraph (c)(1)(ii) of The Rights in Technical Data and Computer Software clause at DFARS 252.227-7013 or subparagraphs (c)(1) and (2) of the Commercial Computer Software — Restricted Rights 48 CFR 52.227-19, as applicable. Manufacturer is Microsoft Corporation, One Microsoft Way, Redmond, WA 98052-6399.

If you acquired this product in the United States, this Agreement is governed by the laws of the State of Washington.

Should you have any questions concerning this Agreement, or if you desire to contact Microsoft Press for any reason, please write: Microsoft Press, One Microsoft Way, Redmond, WA 98052-6399.

Register Today!

Return this

Microsoft® Manual of Style for Technical Publications, Second Edition

registration card for

a Microsoft Press® catalog

U.S. and Canada addresses only. Fill in information below and mail postage-free. Please mail only the bottom half of this page.

1-57231-890-2

MICROSOFT® MANUAL OF STYLE FOR TECHNICAL PUBLICATIONS, SECOND EDITION

Owner Registration Card

NAME

INSTITUTION OR COMPANY NAME

ADDRESS

CITY STATE ZIP

NO POSTAGE
NECESSARY
IF MAILED
IN THE
UNITED STATES

BUSINESS REPLY MAIL
FIRST-CLASS MAIL PERMIT NO. 53 BOTHELL, WA

POSTAGE WILL BE PAID BY ADDRESSEE

MICROSOFT PRESS REGISTRATION
MICROSOFT® MANUAL OF STYLE FOR
TECHNICAL PUBLICATIONS,
SECOND EDITION
PO BOX 3019
BOTHELL WA 98041-9946